Springer Desktop Editions in Chemistry

L. Brandsma, S. F. Vasilevsky, H. D. Verkruijsse
Application of Transition Metal Catalysts in Organic Synthesis
ISBN 3-540-65550-6

H. Driguez, J. Thiem (Eds.)
Glycoscience, Synthesis of Oligosaccharides and Glycoconjugates
ISBN 3-540-65557-3

H. Driguez, J. Thiem (Eds.)
Glycoscience, Synthesis of Substrate Analogs and Mimetics
ISBN 3-540-65546-8

H. A. O. Hill, P. J. Sadler, A. J. Thomson (Eds.)
Metal Sites in Proteins and Models, Iron Centres
ISBN 3-540-65552-2

H. A. O. Hill, P. J. Sadler, A. J. Thomson (Eds.)
Metal Sites in Proteins and Models, Phosphatases, Lewis Acids and Vanadium
ISBN 3-540-65553-0

H. A. O. Hill, P. J. Sadler, A. J. Thomson (Eds.)
Metal Sites in Proteins and Models, Redox Centres
ISBN 3-540-65556-5

A. Manz, H. Becker (Eds.)
Microsystem Technology in Chemistry and Life Sciences
ISBN 3-540-65555-7

P. Metz (Ed.)
Stereoselective Heterocyclic Synthesis
ISBN 3-540-65554-9

H. Pasch, B. Trathnigg
HPLC of Polymers
ISBN 3-540-65551-4

T. Scheper (Ed.)
New Enzymes for Organic Synthesis, Screening, Supply and Engineering
ISBN 3-540-65549-2

Springer
*Berlin
Heidelberg
New York
Barcelona
Hong Kong
London
Milan
Paris
Singapore
Tokyo*

H. Pasch, B. Trathnigg

HPLC of Polymers

Springer

Priv.-Doz.
Dr. Harald Pasch
Deutsches Kunststoff-Institut
Schloßgartenstraße 6
D-64289 Darmstadt
Germany

Prof. Dr. Bernd Trathnigg
Institut für Organische Chemie
Karl-Franzens-Universität
Heinrichstraße 28
A-8010 Graz
Austria

Description of the Series

The Springer Desktop Editions in Chemistry is a paperback series that offers selected thematic volumes from Springer chemistry review series to graduate students and individual scientists in industry and academia at very affordable prices. Each volume presents an area of high current interest to a broad non-specialist audience, starting at the graduate student level.

Formerly published as hardcover edition Springer Laboratory ISBN 3-540-61689-6

Cataloging-in-Publication Data applied for

ISBN 3-540-65551-4
Springer-Verlag Berlin Heidelberg New York

The use of general descriptive names, registered names, trademarks, etc. in this publication does not imply, even in the absence of a specific statement, that such names are exempt from the relevant protective laws and regulations and therefore free for general use.

Cover: design & production, Heidelberg
Typesetting: Data-conversion by Medio, V. Leins, Berlin
SPIN: 10711988   02/3020 - 5 4 3 2 1 0 - Printed on acid-free paper

Preface

Liquid chromatography is one of the workhorses in the analysis of polymers. When it comes to the determination of the molar mass distribution, there is no other technique that can compare with size exclusion chromatography in terms of accuracy and reliability. However, size exclusion chromatography separates according to the size of the macromolecules and not molar mass, and has its limits when very complex polymer systems must be analysed.

The development of polymer systems such as random, block and graft copolymers, polymer blends, telechelics and macromonomers necessitated liquid chromatography to be used not only for molar mass determinations but also for chemical heterogeneity and functionality type distribution. This became possible after the introduction of a number of techniques that are summarized in the term interaction chromatography. These include liquid adsorption chromatography, as well as gradient elution and precipitation chromatography.

Last but not least, liquid chromatography at the critical point of adsorpion, developed by the Russian groups of Belenkii in St. Petersburg and Entelis in Moscow, has brought important new facets into liquid chromatography of polymers.

A number of textbooks on liquid chromatography of polymers have been published in recent years, covering the fundamentals of the different techniques. However, to our knowledge there is no edition on the market that describes the experimental aspect of the different techniques, in particular in interaction chromatography, and that gives detailed instructions for conducting experiments using the diverse techniques. Since experiment is always the proof of the theory, the authors intended to give an introduction into liquid chromatography by proposing a number of more or less simple experiments.

This laboratory manual is written as well for beginners as for experienced chromatographers. The subject of the book is the description of the experimental approach to the analysis of complex polymers. It summarizes important applications in size exclusion and liquid adsorption chromatography and covers the development of separations using liquid chromatography at the critical point of adsorption. The theoretical background, equipment, experimental procedures and applications are discussed for each separation technique. This will enable polymer chemists, physicists and material scientists, as well as students of macromolecular and analytical sciences to optimize experimental conditions for a specific separation problem. The main benefit to the reader is that a wide variety of instrumentation, separation procedures, and applications is presented, making it possible to solve simple as well as sophisticated separation tasks.

The authors wish to express their gratitude and appreciation to all their collegues who provided experimental details on specific applications in liquid chromatography, in particular H. Much and G. Schulz, (Berlin), J. Heyne and M. Augenstein, (Darmstadt), and S. Mori, (Tsu). The authors thank the editors of this book, G. Glöckner, (Dresden), and H. Barth, (Wilmington), for their continuous support. Carefully rewieving a book means lots of work and not much appreciation for the reviewers. Therefore, the authors wish to express their deep gratitude to S. Balke, (Toronto), H. Barth, (Wilmington), and P. Kilz, (Mainz), for performing this difficult task and reading the manuscript carefully.

Darmstadt, October 1997   Harald Pasch

Graz, October 1997   Bernd Trathnigg

Table of Contents

Symbols and Abbreviations

a	Mark-Houwink exponent
ACN	acetonitrile
AH	adipic acid-hexane diol polyester
c	concentration
CC	critical chromatography
CCD	chemical composition distribution
DAD	diode-array detector
D-RI	dual density-refractive index detection
DVB	divinyl benzene
e	grafting success
ELSD	evaporative light scattering detector
EPDM	ethylene-propylene-diene rubber
f	molecular functionality
f_n	number-average functionality
f_p	practical functionality
f_w	weight-average functionality
FAE	fatty alcohol ethoxylate
FTD	functionality type distribution
G	molar ratio in a monomer mixture
g	grafting degree
ΔG	Gibbs free energy
GPC	gel permeation chromatography
GTP	group transfer polymerization
H	height equivalent of a theoretical plate
h	grafting height
ΔH	interaction enthalpy
ΔH_m	mixing enthalpy
HEMA	hydroxyethyl methacrylate
HPLC	high performance liquid chromatography
HPPLC	high performance precipitation liquid chromatography
IR	infrared
K	constant factor in the Mark-Houwink equation
K*	optical constant in light scattering
K_d	distribution coefficient
K_{LAC}	distribution coefficient of adsorption
K_{Rep}	distribution coefficient of repulsion
K_{SEC}	distribution coefficient of size exclusion
l	bond distance

L_A	sequence length of sequence A
L_B	sequence length of sequence B
LAC	liquid adsorption chromatography
LALLS	low angle light scattering
LCCC	liquid chromatography at critical conditions
M	molar mass
M_{eq}	equivalent molar mass
M_n	number-average molar mass
M_0	molar mass of repeat unit
M_V	viscosity-average molar mass
M_W	weight-average molar mass
MALDI-MS	matrix-assisted laser desorption/ionization mass spectrometry
MALLS	multi angle light scattering
MEK	methyl ethyl ketone
MeOH	methanol
MMA	methyl methacrylate
MMD	molar mass distribution
m_i	mass of species i
N	plate number
n	degree of polymerization
n_i	number of species i
NMR	nuclear magnetic resonance
NP	normal phase
P	degree of polymerization
P_n	number-average degree of polymerization
P_W	weight-average degree of polymerization
PEG	polyethylene glycol
PEO	polyethylene oxide
PnBMA	poly-n-butyl methacrylate
PtBMA	poly-t-butyl methacrylate
PDMA	polydecyl methacrylate
PMMA	polymethyl methacrylate
poly-THF	polytetrahydrofuran
PPG	polypropylene glycol
PPO	polypropylene oxide
PS	polystyrene
PTO	polytrioxocane
r_A	reactivity ratio of monomer A
r_B	reactivity ratio of monomer B
$R (\Theta)$	intensity of scattered light
RALLS	right angle light scattering
RI	refractive index
RP	reversed phase
RT	retention time
ΔS	conformational entropy
ΔS_m	mixing entropy

S-EA	styrene-ethyl acrylate copolymer
SEC	size exclusion chromatography
SFC	supercritical fluid chromatography
St	styrene
T	temperature
TALLS	two angle light scattering
THF	tetrahydrofuran
U	polydispersity
UV	ultraviolet
V	packing volume
v	flow velocity
V_e	elution volume
V_i	interstitial volume
V_P	pore volume
V_R	retention volume
V_{stat}	volume of the stationary phase
VISC	viscometer
w	weight fraction
ε	potential of interaction
ε^0	solvent strength
ε_c	finite critical energy of adsorption
η	viscosity of a solution
$[\eta]$	intrinsic viscosity, Staudinger index
η_0	viscosity of a solvent
η_{rel}	relative viscosity
η_{sp}	specific viscosity
λ	wavelength

1 Introduction

1.1 The Molecular Structure of Polymers

Polymers are highly complex multicomponent materials. Properties typically considered important to polymer performance in products may be very diverse and can be divided into simple and distributed properties. Simple properties are the total weight of polymer present, the residual monomer or oligomer content, total weight of microgels or aggregates, and properties that depend only on these measures, such as conversion in the polymerization reaction, monomer composition and average copolymer composition. For other properties different molecules in the same polymer will have different values of the property. These properties are termed distributed properties, the most important of them in polymer chemistry being the molar mass distribution, the distribution of compositions, the distribution of sequence lengths and the distribution of functionality.

The breadth of the distribution of a distributed property determines whether the polymer is homogeneous or heterogeneous with respect to that property. If a property distribution is "broad" then this means that many different values of that property are present in the polymer and the polymer is considered heterogeneous with respect to the property. Accordingly, a polymer is homogeneous or has "narrow" distribution of in a certain property when it has low variety in values of that property. A polymer with more than one type of distributed property can be simultaneously heterogeneous in all these properties. However, it may occur that a polymer is heterogeneous in one or more distributed properties while being homogeneous in other distributed properties. For example, a functional copolymer can be distributed in molar mass and composition but homogeneous in functionality. Using these concepts, monodisperse, polydisperse, and complex polymers can be defined as follows [1]:

- Monodisperse polymers are homogeneous in all distributed properties (for example a homopolymer sample that has a narrow molar mass distribution).
- Polydisperse polymers are heterogeneous in one, and only one, distributed property (for example a homopolymer sample that has a broad molar mass distribution).

– Complex polymers are heterogeneous in more than one distributed property (for example linear copolymers are distributed in molar mass and composition).

In general, the molecular structure of a macromolecule is described by its size, its chemical structure, and its architecture. The chemical structure characterizes the constitution of the macromolecule, its configuration and its conformation. For a complete description of the constitution the chemical composition of the polymer chain and the chain ends must be known. In addition to the type and quantity of the repeat units their sequence of incorporation must be described (alternating, random, or block in the case of copolymers). Macromolecules of the same chemical composition can still have different constitutions due to constitutional isomerism (1,2- vs 1,4-coupling of butadiene, head-to-tail vs head-to-head coupling, linear vs branched molecules). Configurational isomers have the same constitution but different steric patterns (*cis*- vs *trans*-configuration; isotactic, syndiotactic and atactic sequences in a polymer chain). Conformational heterogeneity is the result of the ability of fragments of the polymer chain to rotate around single bonds. Depending on the size of these fragments, interactions between different fragments, and a certain energy barrier, more or less stable conformations may be obtained for the same macromolecule (rodlike vs coil conformation).

Depending on the composition of the monomer feed and the polymerization procedure, different types of heterogeneities may become important. For example, in the synthesis of tailor-made polymers frequently telechelics or macromonomers are used. These oligomers or polymers usually contain functional groups at the polymer chain end. Depending on the preparation procedure, they can have a different number of functional endgroups, i.e. be mono-, bifunctional etc. In addition, polymers can have different architectures, i.e. they can be branched (star- or comblike), and they can be cyclic.

From the point of view of polymer analysis the different heterogeneities can be summarized in the term "molecular heterogeneity", meaning the different aspects of molar mass distribution, distribution in chemical composition, functionality type distribution and molecular architecture. They can be superimposed one on another, i.e. bifunctional molecules can be linear or branched, linear molecules can be mono- or bifunctional, copolymers can be block or graft copolymers etc., see Fig. 1.1. In order to characterize complex polymers, i.e. polymers which are distributed in more than one property, it is necessary to know the molar mass distribution within each other type of heterogeneity.

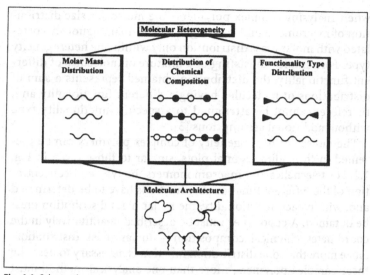

Fig. 1.1 Schematic representation of the molecular heterogeneity of polymers

Clearly, it is very difficult in a general case to solve this characterization problem.

Using the traditional methods of polymer analysis, such as infrared spectroscopy or nuclear magnetic resonance (NMR), one can determine the type of monomers or functional groups present in the sample. However, the determination of functional groups is complicated for long molecules because of low concentration. On the other hand, these methods do not yield information on how functional groups or different monomer species are distributed in the polymer molecule. The conformation and the configuration of a polymer chain may be determined by spectroscopic and X-ray methods, by investigating the solubility or the melting behaviour. But again, average numbers instead of distributions are obtained.

With respect to methods sensitive to the size of the macromolecule, one can encounter other difficulties. Although there is a multitude of methods for molar mass determination, some of these methods are not absolute but relative methods. Further, depending on the molar mass range, different methods must be used, such as vapour pressure osmometry and end group determination for the lower molar mass range, and light scattering, ultracentrifugation and membrane osmometry for the higher molar mass range [2].

Size exclusion chromatography, which is most frequently used to separate polymer molecules from each other according to their molecular size in solution, must be used very carefully when analysing complex polymers. The molecular size distribution of macromolecules can in general be unambiguously correlated with molar mass distribution only within one heterogeneity type. For samples consisting of a mixture of molecules of different functionality, the distribution obtained represents a sum of distributions of molecules having a different functionality and, therefore, cannot be attributed to a specific functionality type without additional assumptions [3].

The molecular heterogeneity of complex polymers can be presented in three-dimensional plots, similar to those given in Fig. 1.2. For telechelics and macromonomers the type and concentration of the different functionality fractions have to be determined and, within each functionality, the molar mass distribution must be obtained. A copolymer must be described quantitatively in the coordinates chemical composition – molar mass distribution. Since more than one distribution function is necessary to describe these complex polymers, more than one analytical method must be used. It is most desirable that each method used is sensitive towards a specific type of heterogeneity. Maximum efficiency can

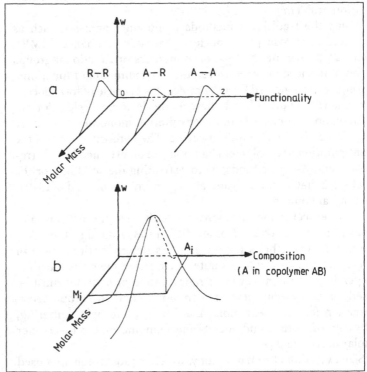

Fig. 1.2. Representation of the molecular heterogeneity of a telechelic polymer (a) and a copolymer (b)

be expected when, similar to the two-dimensional distribution in properties, two-dimensional analytical techniques are used. A possible approach in this respect is the coupling of different chromatographic modes in two-dimensional chromatography or the coupling of a separation technique with selective detectors.

1.2 Molar Mass Distribution

The characteristic feature of a molecule is its molar mass. Low molar mass compounds and most biopolymers are monodisperse with respect to molar mass, i.e. they have a strictly defined molar mass, which may be calculated knowing the chemical composition of the molecule.

The degree of polymerization of a macromolecule P is defined as the number of repeat units within this molecule. P is related to molar mass M by

$$P = M / M_0 \tag{1.1}$$

where M_0 represents the molar mass of the repeat unit.

All synthetic polymers are polydisperse or heterogeneous in molar mass. The *molar mass distribution* originates from randomness of the polymerization process. In the daily routine synthetic polymers are often characterized by average molar masses, considering the frequencies (numbers) of macromolecules of a certain molar mass M_i in the sample. Most frequently used are the *number-average* molar mass M_n, expressing the amount of species by the number of moles n_i, and the *weight-average* molar mass M_w, considering the mass m_i of the species. As m_i is related to n_i via $m_i = n_i M_i$, and for a single species $M_i = M_0 P_i$, the molar mass averages may be expressed as average degrees of polymerization, where P_n is the number-average and P_w is the weight-average degree of polymerization.

$$M_n = \Sigma n_i M_i / \Sigma n_i = P_n M_0 \tag{1.2}$$

$$M_w = \Sigma w_i M_i / \Sigma w_i = \Sigma n_i M_i^2 / \Sigma n_i M_i = P_w M_0 \tag{1.3}$$

Molar masses of polymers may be determined by different methods (see Table 1.1) [4]. The difference between number and weight average molar masses gives a first estimate of the width of the molar mass distribution (MMD). The broader the distribution, the larger is the difference between M_n and M_w. The ratio of M_w/M_n is a measure of the breadth of the molar mass and is termed the polydispersity U

Table 1.1. Principle methods of determining molar masses

Method	Type	Molar Mass Range (g/mol)	Average Value
membrane osmometry	A	$10^4 - 10^5$	M_n
cryoscopy	A	$< 10^4$	M_n
vapor pressure osmometry	A	$< 10^4$	M_n
end-group analysis	E	$< 10^4$	M_n
light scattering	A	$10^3 - 10^7$	M_w
ultracentrifugation	A	$10^3 - 10^7$	different averages
viscometry	R	$10^2 - 10^7$	M_v
size exclusion chromatography	R	$10^2 - 10^7$	different averages

$$U = M_w/M_n - 1 \qquad\qquad (1.4)$$

For absolute methods (A), the molar mass is directly calculated from the experimental data, without additional information on the chemical structure of the polymer. Equivalent methods (E), in particular molar mass determination by end-group analysis, requires that the chemical structure of the end group is known. Relative methods (R) are based on the physical behaviour of the polymer and the interactions with a solvent. In this case the method must be calibrated with samples of known molar mass.

The most important method for determining the molar mass distribution is size exclusion chromatography (SEC), also referred to as gel permeation chromatography (GPC) [5–7]. SEC separates polymers with respect to their hydrodynamic volume. For homopolymers the hydrodynamic volume is directly related to the molar mass, and using a calibration for the polymer under investigation the molar mass averages and the polydispersity can be determined. Applying SEC to the analysis of copolymers is more difficult, since hydrodynamic volume is a function of molar mass and chemical composition. In this case more sophisticated techniques must be used, such as [1, 8]: conventional SEC utilizing spectroscopic detection, on-line laser light scattering detection, on-line viscometry, two-dimensional HPLC-SEC separation.

1.3 Chemical Composition Distribution

When two or more monomers of different chemical structures are involved in a polymerization reaction, instead of a chemically homogeneous homopolymer in most cases a chemically heterogeneous copolymer is formed. Depending on the reactivity of the monomers and their sequence of incorporation into the polymer chain, macromolecules can be formed which differ significantly in composition, (meaning the amounts of repeat units A, B etc. in the copolymer) and the sequence distribution. With respect to sequence distribution copolymers can be classified as alternating, random, block and graft copolymers.

Chemical heterogeneity is a consequence of CCD and can be presented as an integral or differential distribution curve of composition vs molar mass. Information on the complex MMD/CCD can be provided as shown in Fig. 1.3 for a bipolymer, plotting mole fraction of A or B units vs molar mass of the copolymer.

Consider a random copolymer which has been obtained in a homogeneous reaction from a mixture of A and B monomers. Even under such favourable conditions the resulting macromolecules will differ in chemical structure. There are differences in the sequence of the A and B monomers along the macromolecules, differences in the average chemical composition of the copolymer molecules formed at any instant of the polymerization (instantaneous heterogeneity), and differences due to the depletion of the reaction mixture in one of the monomers.

The sequence distribution of a copolymer chain may be characterized by the number-average lengths of uninterrupted

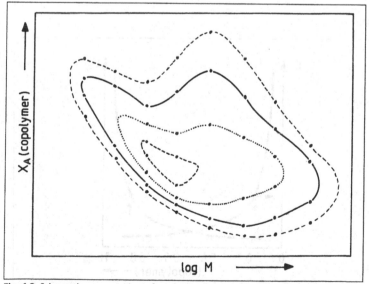

Fig. 1.3 Schematic presentation of a contour-line map fraction of A units in a random copolymer vs molar mass of the copolymer; the contour lines connect points of equal height corresponding to equal concentration

sequences of A and B units in this chain. The sequence lengths L_A and L_B can be estimated from the copolymerization propagation probabilities or the monomer reactivity ratios r_A and r_B.

$$L_A = 1 + r_A\, G \tag{1.5}$$

$$L_B = 1 + r_B\, G \tag{1.6}$$

where G is the molar ratio [A]/[B] in the monomer mixture. Figure 1.4 is a schematic representation of a plot of average sequence length vs mole fraction of monomer A in the copolymer.

The average sequence lengths L_A and L_B can be measured by physical or chemical methods. The former (FTIR or NMR analyses) usually measure the percentage of A and B units inside of triads, pentads etc. whereas the latter methods evaluate the percentage of A-A, A-B and B-B linkages. Macromolecules of random copolymers, even if identical in chain length and composition (and thus also in the average sequence length) still offer a great variety with respect to the order of individual sequences in the molecules. Thus, a copolymer sample contains a tremendous number of constituents. In terms of liquid chromatography, a sample of this kind is an extremely complex mixture, difficult to separate by size exclusion or interaction chromatography [9].

In addition to the sequence distribution, conversion heterogeneity has to be considered when analysing copolymers. Only in special cases is the composition of a copolymer identical with the

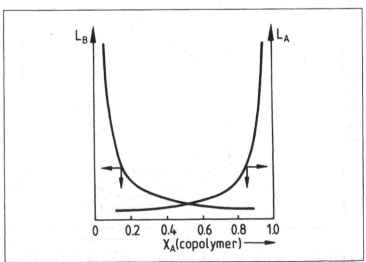

Fig. 1.4. Schematic representation of average sequence lengths L_A and L_B versus copolymer composition

composition of the monomer batch. These cases are azeotropic copolymers or systems whose monomer reactivity ratios equal 1. In general, the instantaneous composition of a copolymer differs from the composition of the monomer mixture, which causes depletion in the batch of the monomer which is preferably incorporated. Thus, subsequent portions of a copolymer sample are polymerized from mixtures of various compositions and this gives rise to additional chemical heterogeneity. Accordingly, when discussing CCD of copolymers, *sequence distribution*, *instantaneous heterogeneity* and *conversion heterogeneity* must be considered.

1.4 Functionality Type Distribution

Oligomers and polymers with reactive functional groups have been used extensively to prepare a great variety of polymeric materials. In many cases, the behaviour and reactivity of these functional homopolymers is largely dependent on the nature and the number of functional groups. In a number of important applications the functional groups are located at the end of the polymer chain. Macromolecules with terminal functional groups usually are termed "telechelics" or "macromonomers".

Molecular functionality, f, of a telechelic polymer is described as the number of functional groups per molecule. If some of these functional groups are not reactive, the practical functionality, f_p, is lower than the molecular functionality, $f_p<f$.

Macromolecules with the same structure of the polymer chain may be different in the number and the nature of the functional groups. Depending on the number of new bonds they can form in a reaction, one can classify functional groups as single-act groups, forming one new bond (-OH, -COOH, $-NH_2$, -COCl), and dual-act or multi-act groups, forming two or more new bonds (C=C, -CO-O-CO-, -N=C=O). To prepare a linear polymer by polymerization or polycondensation, each molecule must have f=2 for single-act or f=1 for dual-act groups. To obtain a cross-linked polymer, functionality must be f>3 or f>2, respectively.

When functional homopolymers are synthesized, functionally defective molecules are formed in addition to macromolecules of required functionality (see Fig. 1.5). For example, if a target functionality of f=2 is required, then in the normal case species with f=1, f=0 or higher functionalities are formed as well [10], which may result in a decreased or increased reactivity, cross-linking density, surface activity etc. Each functionality fraction has its own molar mass distribution. Therefore, for complete

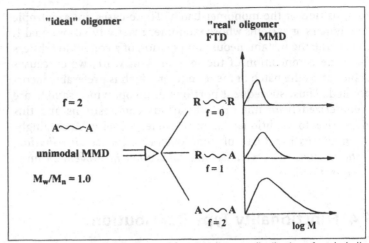

Fig. 1.5. Molar mass distribution and functionality type distribution of a telechelic oligomer (Reprinted from Ref. [3] with permission of Springer-Verlag)

description of the chemical structure of a functional homopolymer, the determination of the molar mass distribution and the functionality type distribution is required.

Typically, functionality is quantitatively described as a number-average functionality, f_n, where f_n is the ratio of the total number of functional groups to the total number of molecules in the system, i.e. the average number of functional groups per initial molecule. It is experimentally determined from

$$f_n = M_n/M_{eq} \qquad (1.7)$$

where M_n is the number-average molar mass and M_{eq} is the equivalent molar mass, that is the average mass of the molecule per one functional group.

The f_n value provides information on the average functionality but does not characterize the functional polydispersity. An average functionality of 1 may be simulated by equal amounts of non-functional and two-functional species, and is therefore ambiguous. The characterization of the width of the functionality type distribution is more informative. In analogy to the average molar masses, number-average and weight-average functionalities may be introduced,

$$f_n = \Sigma n_i f_i / \Sigma n_i \qquad (1.8)$$

$$f_w = \Sigma w_i f_i / \Sigma w_i = \Sigma n_i f_i^2 / \Sigma n_i f_i \qquad (1.9)$$

where n_i is the number of molecules of functionality f_i, and $w_i = n_i f_i$.

Table 1.2. Classification of functional homopolymers

Type	General Structure	Polydispersity Parameters
1	A∿∿∿R A∿∿∿A A∿∿∿A with A branch	$M_w/M_n \geq 1$ $f_w/f_n=1$
2	[∿∿∿]$_n$ with A branch	$f_w/f_n \sim M_w/M_n$
3	A branch, A∿∿∿A	$f_w/f_n \geq 1$ $M_w/M_n \geq 1$

For the description of the functional polydispersity the term f_w/f_n may be used. For polymers containing only one type of molecules, $f_w/f_n=1$ is obtained; in the case of a distribution of molecules of different functionality $f_w/f_n > 1$ is obtained.

The simultaneous use of the functionality type distribution (FTD) and MMD to characterize functional homopolymers enables subdivision of the systems into three basic types [10]:

- Type 1: Oligomers with a strictly defined target functionality. Ideally these oligomers must have $f_w/f_n=1$. However, in reality a certain amount of functionally defective molecules are formed, introducing a distribution in functionality.
- Type 2: Polyfunctional linear or branched oligomers with a regular alternation of functional groups along the chain.
- Type 3: Polyfunctional linear or branched oligomers with irregular alternation of functional groups in the chain. These oligomers may have diverse values of functional and molar mass polydispersity (see Table 1.2) [10].

1.5 Polymer Solution Properties

Polymer chain structures range from the extreme of rodlike chains or helix structures to chains with a maximum of conformational states due to rotational isomerism about the valence bonds along the chain backbone. In macromolecular science the latter category relates to *flexible chain polymers*. Under certain conditions, interactions among chain elements can be neglected in describing the conformational states of flexible chains. The idealized chain without interactions is called an unperturbed chain, and the mean-square end-to-end distance of the chain $<r^2>_0$ scales with number of repeat units in the chain nl^2, where l is the bond distance of the backbone monomer and n is the number of bonds. In practice, in many cases, the average chain conformation obtained in a dilute solution at a certain temperature approximates that of the unperturbed chain. This temperature, by definition, is the theta temperature at which the second virial coefficient in the osmotic equation vanishes for infinitely long chains.

The conformation of a flexible unperturbed polymer chain can be described using different models. The "random-flight chain" model is modelled by the path of a Brownian particle [11]. For semiflexible or stiff-chain structures, the persistent or "wormlike"-chain model may be utilized [12]. The wormlike-chain model affords a continuous description of chain character ranging from flexible to rodlike.

The dimension of a linear-chain molecule is usually expressed in terms of the end-to-end distance r. For a given structure of the basic chain, the mean-square value of r is determined by the nature of the hindrance to internal rotation around single bonds and by van der Waals or other types of interactions between non-bonded groups that are separated in the basic chain structure by many valence bonds. This latter interaction is called a long-range interaction, while the hindrance to internal rotation is called a short range interaction.

In the absence of both types of interactions, the so called "freely rotating chain" is obtained, and its dimension is easily computed from bond lengths and angles $<r^2>_0 = nl^2$. The dimension of a chain lacking long-range interactions and consisting of only one kind of bond is expressed as the mean-square value of the end-to-end distance $<r^2>_f$

$$<r^2>_f = n\, l^2 \left[(1 + \cos\theta)/(1 - \cos\theta)\right] \qquad (1.10)$$

where θ is the supplement of the valence bond angle, and the subscript f denotes the freely rotating state lacking long-range interactions.

The hindrance of internal rotations introduces appreciable changes in the average chain dimensions, but it does not alter the proportionality between r and n. Therefore a chain without short- and long-range interactions is an unperturbed chain, and its dimension is the unperturbed dimension. The ratio of the unperturbed dimension r_o to r_f represents the effect of steric hindrance, in which the value of σ is independent of n.

$$\sigma = r_o/r_f = [<r^2>_o/<r^2>_f]^{0.5} \qquad (1.11)$$

Long-range interactions give rise to the "excluded-volume effect" which can be described as an osmotic swelling of the randomly coiled chain by solvent-polymer interactions. As a result of the superposition of short- and long-range interactions, the average end-to-end distance of a real linear macromolecule in dilute solution is generally written as

$$r \equiv <r^2>^{0.5} = \alpha\, r_o = (\alpha\sigma)r_{of} \qquad (1.12)$$

in which the linear expansion factor α depends on the number of bonds n. r_{of} relates to the end-to-end distance in the freely rotating state without long- range interactions.

Another measure of the short-range interactions in polymer chains is the persistence length a_p which is defined as the average projection of an infinitely long chain along the direction of its first link. For an unperturbed chain consisting of one kind of bond, a_p can be written as

$$a_p = 1\,[(r_o/2nl^2) + 0.5] \qquad (1.13)$$

Similar relations between a_p and r_o are also obtainable for more complicated chains [13, 14]. Unperturbed dimensions of linear chain polymers which are obtained under various conditions of solvent and temperature are tabulated in Ref. [15].

The *hydrodynamic properties* of dilute solutions of flexible-chain polymers are closely associated with chain dimensions. The viscosity of a dilute polymer solution depends on the nature of polymer and solvent, the concentration of the polymer, its average molar mass and molar mass distribution, the temperature and the rate of deformation [16, 17]. In the following discussion it is assumed that the rate of deformation is so low that its influence can be neglected.

The ratio of the viscosity η of a solution to the viscosity η_o of the pure solvent is called the viscosity ratio or the relative viscosity η_{rel}

$$\eta_{rel} = \eta/\eta_o \qquad (1.14)$$

The relative increase of the viscosity of a polymer solution to that of the solvent is called the specific viscosity η_{sp}

$$\eta_{sp} = (\eta - \eta_o)/\eta_o = \eta_{rel} - 1 \qquad (1.15)$$

This quantity is divided by the concentration c to obtain the reduced viscosity η_{sp}/c which expresses the average contribution of the solute molecules at concentration c to the viscosity. The limiting reduced viscosity or the Staudinger index $[\eta]$ which is also called the *intrinsic viscosity* is the value of the reduced viscosity at infinite dilution, i.e.

$$[\eta] = \lim \eta_{sp}/c = \lim [(\eta - \eta_o)/\eta_o \, c] \qquad (1.16)$$

The viscosity of polymer solutions, especially with high molar mass polymers, is often appreciably dependent on the rate of shear in the range of measurement. The intrinsic viscosity should, therefore, be given as the limiting value of η_{sp}/c not only at infinite dilution but also at a shear rate of zero, or the value of the rate of shear should be specified.

The intrinsic viscosity is a measure of the capacity of a polymer molecule to enhance viscosity, which depends on the size and the shape of the polymer molecule. Within a given series of polymer homologues, $[\eta]$ increases with M; hence it is a measure of M. The empirical Mark-Houwink-Sakurada equation describes the viscosity-molar mass relationship for Gaussian coils:

$$[\eta] = K \, M_a \qquad (1.17)$$

where K and a are coefficients that are characteristic of the polymer, solvent, and temperature. It is well established that for linear, flexible polymers under special conditions of temperature and solvent, which are known as Θ-*conditions* the above equation becomes

$$[\eta]_\Theta = K_\Theta \, M^{0.5} \qquad (1.18)$$

Since Eq. (1.18) is for an ideal chain at Θ-conditions, there is no molar mass dependence. However, since the coefficients K and a determined in usual solvents are valid only within a limited range of M, it is necessary to specify the molar mass range in which they were determined. For thermodynamically good solvents, a lies between 0.5 and 0.75 for flexible-chain polymers. K and a values are tabulated in a number of compilations [18, 19]. The theoretical exponents a of the Mark-Houwink-Sakurada equation are summarized in Table 1.3.

Table 1.3. Theoretical exponents a of the Mark-Houwink-Sakurada equation [18]

Con-formation	Description	a
rod	diameter constant, length prop. to M, no rotational motion (rigid chain)	2
rod	same, but with rotational motion	1.7
coil	linear chain, unperturbed, excluded volume	0.51-0.9
coil	linear chain, unperturbed, no excluded volume (Θ-conditions)	0.5
disc	diameter prop. to M, length constant	0.5
sphere	constant density, nonsolvated or uniformly solvated	0

References

1. BALKE ST (1991) Characterization of Complex Polymers by Size Exclusion Chromatography and High-Performance Liquid Chromatography. In: BARTH HG, MAYS JW (eds) Modern Methods of Polymer Characterization, Chapter 1. Wiley-Interscience, New York
2. ELIAS GH (1984) Macromolecules, Chapter 9. Plenum Press, New York
3. ENTELIS SG, EVREINOV VV, GORSHKOV AV (1986) Adv Polym Sci 76:129
4. SCHRÖDER E, MÜLLER G, ARNDT K-F (1982) Leitfaden der Polymercharakterisierung. Akademie-Verlag, Berlin
5. JANCA J (ed) (1984) Steric Size Exclusion Chromatography of Polymers. Dekker, New York
6. GLÖCKNER G (1987) Polymer Characterization by Liquid Chromatography. Elsevier, Amsterdam
7. BALKE ST (1984) Quantitative Column Liquid Chromatography: A Survey of Chemometric Methods. Elsevier, Amsterdam
8. KILZ P, KRÜGER R -P, MUCH H, SCHULZ G (1995) ACS Adv Chem 247:223
9. GLÖCKNER G (1991) Gradient HPLC of Copolymers and Chromatographic Cross-Fractionation, Chapter 2. Springer, Berlin Heidelberg New York
10. ENTELIS SG, EVREINOV VV, KUZAEV AI (1985) Reactive Oligomers. Khimiya, Moscow

11. KUHN W (1936) Kolloid-Z 76:258, (1939) Kolloid-Z 78:3
12. KRATKY O, POROD G (1949) Rec Trav Chim (Belgium) 68:1106
13. PETERLIN A (1960) J Polymer Sci 47:403
14. HEINE S, KRATKY O, POROD G (1961) Makromol Chem 44–46:682
15. BRANDRUP J, IMMERGUT, EH (Eds) (1989) Polymer Handbook. Wiley, New York
16. YAMAKAWA H (1971) Modern Theory of Polymer Solutions. Harper & Row, New York
17. BOHDANECKY M, KOVAR J (1982) Viscosity of Polymer Solutions. Elsevier, Amsterdam
18. ELIAS HG (1984) Macromolecules. Plenum Press, New York, p 314
19. GLÖCKNER G (1991) Gradient HPLC of Copolymers and Chromatographic Cross-Fractionation. Springer, Berlin Heidelberg New York, pp 28–31

2 Thermodynamics of Polymer Chromatography

2.1 Entropic and Enthalpic Interactions

Any chromatographic process relates to the selective distribution of an analyte between the mobile and the stationary phase of a given chromatographic system. In liquid chromatography the separation process can be described by

$$V_R = V_i + V K_d \qquad (2.1)$$

where V_R is the retention volume of the solute, V_i is the interstitial volume of the column, V is the volume of the packing, i.e. the "stationary" volume, and K_d is the distribution coefficient which is equal to the ratio of the analyte concentration in the stationary phase and in the mobile phase. Note that V can be comprised of the pore volume V_p, surface area V_a, or volume of chemically bonded "stationary phase" on the packing V_{stat}, depending on the separation mode and type of packing.

K_d is related to the change in Gibbs free energy ΔG related to the analyte partitioning between the mobile and the stationary phase [1].

$$\Delta G = \Delta H - T\Delta S = -RT \ln K_d \qquad (2.2)$$

$$K_d = \exp (\Delta S/R - \Delta H/RT) \qquad (2.3)$$

The change in Gibbs free energy may be due to different effects:

1. Inside the pore, which has limited dimensions, the macromolecule cannot occupy all possible conformations and, therefore, the conformational entropy ΔS decreases.
2. When penetrating the pores, the macromolecule may interact with the pore walls resulting in a change in enthalpy ΔH.

Depending on the chromatographic system and the chemical structure of the macromolecule, only entropic or enthalpic interactions or both may be operating. Therefore, in the general case the distribution coefficient may be expressed as

$$K_d = K_{SEC}\, K_{LAC} \qquad (2.4)$$

where K_{SEC} is based on entropic interactions, while K_{LAC} characterizes the enthalpic interactions. Depending on the magnitude of entropic and enthalpic effects, the size exclusion mode (ΔS) or the adsorption mode (ΔH) will be predominant.

2.2 Size Exclusion Mode

In size exclusion chromatography, separation is accomplished with respect to the hydrodynamic volume of the macromolecules. The stationary phase is a swollen gel with a characteristic pore size distribution, and depending on the size of the macromolecules a larger or lesser fraction of the pores is accessible to the macromolecules.

In *ideal SEC*, separation is exclusively directed by conformational changes of the macromolecules and ΔH by definition is zero, thus

$$K_d = K_{SEC} = \exp(\Delta S/R) \qquad (2.5)$$

Since the conformational entropy decreases ($\Delta S < 0$), the distribution coefficient of ideal SEC is $K_{SEC} < 1$. The maximum value, $K_{SEC} = 1$, is related to zero change in conformational entropy, i.e. to a situation where all of the pore volume is accessible to the macromolecules (separation threshold). At $K_{SEC} = 0$, the analyte molecules are too large to penetrate into the pores (exclusion limit). Accordingly, the separation range is $0 < K_{SEC} < 1$.

The retention volume for ideal SEC is

$$V_R = V_i + V_p\, K_d = V_i + V_p\, K_{SEC} \qquad (2.6)$$

If enthalpic effects, due to electrostatic interactions between macromolecules and the pore walls, have to be taken into account, the distribution coefficient K_d of *real SEC* is as follows:

$$K_d = \exp(\Delta S/R - (\Delta H/RT))$$
$$= \exp(\Delta S/R)\exp(-\Delta H/RT) = K_{SEC}\, K_{LAC} \qquad (2.7)$$

In this case, the retention volume is a function of K_{SEC} and K_{LAC}. If electrostatic interactions occur at the outer surface of the stationary phase as well, an additional term $V_{stat}\, K_{LAC}$ has to be accounted for.

The molar mass vs retention volume behaviour in the size exclusion mode is given in Fig. 2.1. The smaller the macromolecules the more pore volume they can penetrate and, accordingly,

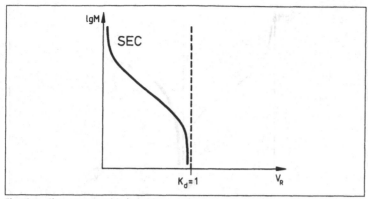

Fig. 2.1. Chromatographic behaviour molar mass vs retention volume in the size exclusion mode

the longer they are retained in the stationary phase. Therefore, first the large macromolecules are eluted followed by macromolecules of smaller size.

2.3 Adsorption Mode

In adsorption chromatography (LAC), where separation is directed by adsorptive interactions between the macromolecules and the stationary phase, an ideal and a real case may be defined as well. In *ideal LAC* conformational changes are assumed to be zero ($\Delta S = 0$) and the distribution coefficient is exclusively determined by enthalpic effects.

$$K_d = K_{LAC} = \exp(-\Delta H/RT) \tag{2.8}$$

Depending on the pore size of the stationary phase two possible cases have to be discussed:

1. For narrow-pore stationary phases, separation occurs exclusively on the outer surface. The pores are not accessible to the macromolecules ($K_{SEC}=0$). Accordingly, the retention volume is a function of the interstitial volume and the volume of the stationary phase (V_{stat})

$$V_R = V_i + V_{stat} K_{LAC} \tag{2.9}$$

2.If the solute can freely penetrate the pore volume of the stationary phase ($K_{SEC}=1$), the pore volume adds to the interstitial volume.

$$V_R = V_i + V_p + V_{stat} K_{LAC} \tag{2.10}$$

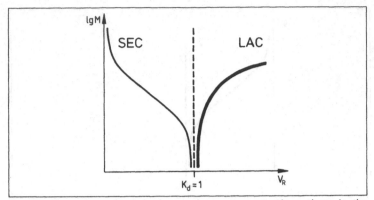

Fig. 2.2. Chromatographic behaviour molar mass vs retention volume in the adsorption mode

In *real LAC* only a fraction of the pores of the packing is accessible and, therefore, entropic interactions must be assumed. Accordingly, the distribution coefficient is a function of ΔH and ΔS (compare to real SEC). The retention volume now is a function of enthalpic interactions at the surface of the packing, entropic effects owing to the limited dimensions of the pores, and possible electrostatic interactions inside the pores. Therefore, the expression for V_R in real LAC is formally similar to that in real SEC

$$V_R = V_i + V_p(K_{SEC} \, K_{LAC}) + V_{stat} \, K_{LAC} \qquad (2.11)$$

The molar mass vs retention volume behaviour in the adsorption mode is shown in Fig. 2.2. As the enthalpic interactions are based on a multiple attachment mechanism, it is clear that the retention volume increases with increasing molar mass [2].

2.4 Critical Mode

Real SEC and real LAC are often mixed-mode chromatographic methods with predominance of entropic or enthalpic interactions. With chemically heterogeneous polymers, effects are even more dramatic because exclusion and adsorption act differently on molecules of different composition.

In a more general sense, the size exclusion mode of liquid chromatography relates to a separation regime where entropic interactions are predominant, i.e., $T\Delta S > \Delta H$. In the reverse case, $\Delta H > T\Delta S$, separation is mainly directed by enthalpic interactions. As both separation modes in the general case are affected by the macromolecule size and the pore size, a certain energy of interaction ε may be introduced, characterizing the specific interactions of the monomer unit of the macromolecule and the stationary phase. ε is a function of the chemical composition of the monomer unit, the composition of the mobile phase of the chromatographic system, the characteristics of the stationary phase and the temperature.

The theory of adsorption at porous adsorbents predicts the existence of a finite critical energy of adsorption ε_c, where the macromolecule starts to adsorb at the stationary phase. Thus, at $\varepsilon > \varepsilon_c$ the macromolecule is adsorbed, whereas at $\varepsilon < \varepsilon_c$ the macromolecule remains unadsorbed. At $\varepsilon = \varepsilon_c$ the transition from the unadsorbed to the adsorbed state takes place, corresponding to a transition from SEC to adsorption. This transition is termed "critical point of adsorption" and relates to a situation where the adsorption forces are exactly compensated by entropy losses [3, 4].

$$T\Delta S = \Delta H \tag{2.12}$$

$$\Delta G = 0 \tag{2.13}$$

Accordingly, at the critical point of adsorption the Gibbs free energy is constant and the distribution coefficient is $K_d=1$, irrespective of the molar mass of the macromolecules and the pore size of the stationary phase. The molar mass vs retention volume behaviour in the *critical mode of liquid chromatography* is schematically represented in Fig. 2.3. The critical point of adsorption relates to a very narrow range between the size exclusion and adsorption modes of liquid chromatography, a region, very sensitive towards temperature and mobile phase composition.

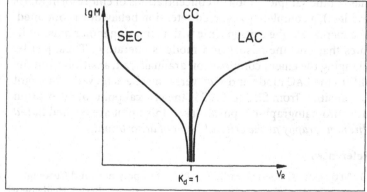

Fig. 2.3. Chromatographic behaviour molar mass vs retention volume in the critical mode

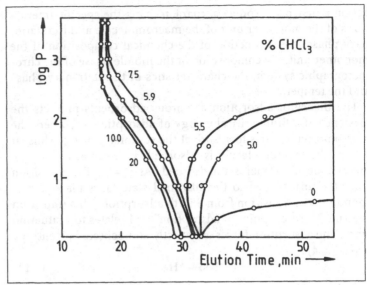

Fig. 2.4. Elution of polystyrene as a function of the eluent composition, stationary phase: silica gel, mobile phase: chloroform-carbon tetrachloride (Reprinted from Ref. [5] with permission)

The transition from one to another chromatographic separation mode by changing the temperature or the composition of the mobile phase for the first time was reported by Tennikov et al. [5] and Belenkii et al. [6, 7]. They showed that a sudden change in elution behaviour may occur by small variations in the solvent strength, see Fig. 2.4, for the behaviour of polystyrene in a mobile phase of chloroform-carbon tetrachloride. In eluents comprising more than 5.9% chloroform, conventional SEC behaviour is obtained, that is elution time increases with decreasing molar mass of the sample. At lower concentrations of chloroform (5.5% and less), a completely reversed retention behaviour is obtained. The increase of the elution time with increasing molar mass indicates that now the adsorption mode is operating. Thus, just by changing the eluent composition gradually, a transition from the SEC to the LAC mode and vice versa may be achieved. The point of transition from SEC to LAC is the critical point of adsorption and chromatographic separations at this point are termed *liquid chromatography at the critical point of adsorption*.

References

1. GLÖCKNER G (1991) Gradient HPLC of Copolymers and Chromatographic Cross-fractionation, chapter 3. Springer, Berlin Heidelberg New York

2. GLÖCKNER G (1987) Polymer Characterization by Liquid Chromatography. Elsevier, Amsterdam
3. ENTELIS SG, EVREINOV VV, GORSHKOV AV (1986) Adv Polym Sci 76: 129
4. ENTELIS SG, EVREINOV VV, KUZAEV AI (1985) Reactive Oligomers. Khimiya, Moscow
5. TENNIKOV MB, NEFEDOV PP, LAZAREVA MA, FRENKEL S J (1977) Vysokomol Soedin A19: 657
6. BELENKII BG, GANKINA ES, TENNIKOV MB, VILENCHIK LZ (1976) Dokl Acad Nauk USSR 231: 1147
7. Skvortsov AM , Belenkii BG , Gankina ES , Tennikov MB (1978) Vysokomol Soedin A20: 678

3 Equipment and Materials

3.1 Solvent Delivery

The following types of pumps are typically used in HPLC (Fig. 3.1):

1. **Syringe pumps** work like a large syringe, the plunger of which is actuated by a screw-feed drive (usually by a stepper motor), hence they deliver a completely pulseless flow
2. **Reciprocating pumps** exist in various modifications:
   a) **Single piston pumps** are cheap, but in general not well suited for SEC.
   b) **Dual piston pumps** with parallel pistons deliver a smooth flow.
   c) **Dual piston pumps** with pistons in series are easier to maintain, because they have only two check valves instead of four. The slightly higher pulsations, however, can be reduced by a pulse dampener to a level comparable to that of the parallel arrangement.

When selecting the pump for a given separation, one has to decide according to the separation technique: positive displace-

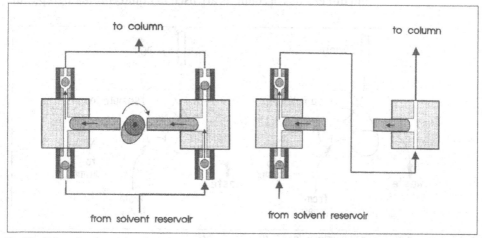

Fig. 3.1. Types of reciprocating dual piston pumps: pistons parallel *(left)* and in series *(right)*

ment pumps offer considerable advantages in SEC (flow stability) and LCCC (no evaporation of solvent, and thus absolutely constant composition of the mobile phase), but cause problems, when the mobile phase has to be changed. In gradient elution, only high-pressure mixing is possible with this type of pump, which makes the whole system even more expensive, because it requires two pumps. On the other hand, reciprocating pumps allow gradient elution also with low-pressure mixing (i.e with just one pump).

3.2 Injection Systems

Typically, two-position six-port valves are used for sample injection, which may be operated manually or automatically (such as in an autosampler). The principle is shown schematically in Fig. 3.2.

For high precision, the loop should always be filled completely with the sample, and a sufficient amount of the sample solution (at least three volumes of the loop) should be used to rinse it thoroughly. The optimum size of the sample loop is determined by the column dimensions, the sensitivity of the detector(s), and by the nature of the separation: in SEC of high molar mass samples, it is recommended to use a larger loop (50–100 µL) rather than a higher concentration of the sample solution. In LAC and LCCC, higher concentrations may be injected, and a smaller loop (10–50 µL) is preferred.

3.3 Columns

Depending on the mechanism of the separation, different types and dimensions of columns are used in liquid chromatography of

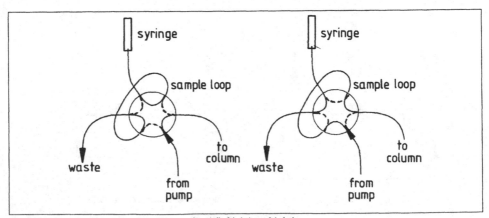

Fig. 3.2. Function of LC injection systems, load *(left)*, inject *(right)*

polymers. As SEC columns will be discussed in detail in Chapter 4, this section will focus on columns for adsorption chromatography and liquid chromatography under critical conditions.

3.3.1 Dimensions

Unlike in SEC, where high separation efficiency can only be achieved with long columns, there is nowadays a trend towards smaller columns in HPLC. As retention is determined by the distribution coefficients between stationary phase and mobile phase, the composition of the latter is the parameter governing the separation. Smaller columns mean faster analyses as well as lower solvent consumption. In analytical applications, the diameter and also the length of columns may be reduced. In practice, there are, however, limitations of miniaturization:

- The quality of the column packing (the plate height) determines the length of the column, which is required to achieve the desired plate number and resolution.
- High efficiency (a lower plate height) can only be achieved with smaller particles, which can be packed more densely. Hence a higher back pressure will result for such a column, the magnitude of which is determined by the flow rate, the viscosity of the mobile phase, and the diameter of the column.
- The length and diameter of the connecting capillaries and the internal volume of the detector cell have also to be small to maintain the overall efficiency of the system.

A classification according to column dimensions is shown in Table 3.1.

Microbore LC requires special injection systems, detectors and capillaries, while narrow bore systems can be operated with normal equipment – at least in principle. In general, the entire system must be assembled very carefully for small column dimensions.

Table 3.1 Classification of HPLC columns with respect to size

Type	Column diameter (mm)	Column length (mm)	Particle size (μm)
Microbore	1–2	5	3
Narrow bore	2–3	5–10	3
Analytical	4–5	10–25	3–5
Semipreparative	<10	25	5–10
Preparative	>10	>25	>10

3.3.2 Stationary Phases

Basically, HPLC columns can be packed with different stationary phases, the nature of which is determined by the separation problem. As a general rule, the efficiency of a column increases with decreasing particle size, but, as an undesirable side effect, so does the back pressure at a given flow rate and given mobile phase viscosity. Spherical particles are generally superior to particles with an irregular shape.

Porous particles are used in SEC where the separation occurs exclusively by limited accessibility of the pores (see Chapter 4), and LAC because of their higher surface area. In critical chromatography (CC) (see Chapter 6), where both effects (exclusion and adsorption) compensate each other, the pore size distribution is also highly important. A stationary phase with small pores will exclude high molar masses, hence no critical conditions will exist beyond this limit. Recently, non-porous phases with a particle diameter of 1.5 μm have been introduced, which may reduce analysis time and solvent consumption in LAC dramatically. These packings are available in columns of 3.3–5 cm length and a diameter of 4.6 mm.

From a chemical point of view, most stationary phases are based on silica or (mainly in SEC) on crosslinked organic polymers. For special applications, alumina can also be used as a matrix. A common classification refers to the polarity of stationary and mobile phase: in normal-phase separations the stationary phase is more polar than the mobile phase, and the opposite is true for reversed-phase LC. Some typical polar stationary phases, as they are used in normal-phase or in ion-exchange LC, are shown in Fig. 3.3. Some typical non-polar stationary phases are given in Fig. 3.4.

In reversed-phase LC, stationary phases based on modified silica are most frequently used, which are typically obtained by reacting them with silanes. As this modification is seldom complete, the residual silanol groups may affect the separation, especially in the analysis of basic compounds, such as amines. Most producers offer packings with high carbon load and a high degree of end-capping, which should minimize these effects. The safest way, however, is the use of polymer-based packings (mainly crosslinked styrene-divinylbenzene copolymers). Note that the nitrile phase can be used either in normal or in reversed-phase chromatography, depending on the polarity of the mobile phase.

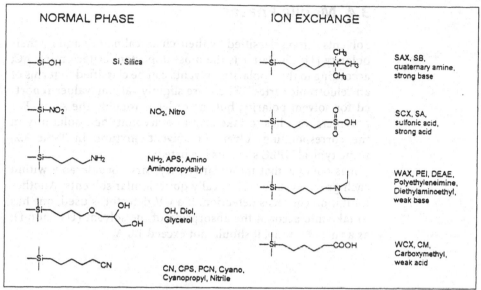

Fig. 3.3. Typical polar stationary phases for liquid chromatography

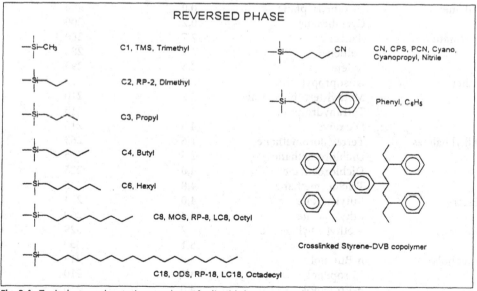

Fig. 3.4. Typical non-polar stationary phases for liquid chromatography

3.4 Mobile Phases

Solvents can be classified by their chemical nature and by their polarity [1]. The latter is the most important criterion in LAC; according to their polarity, solvents can be classified in terms of an "eluotropic series". There are slightly varying values reported for solvent polarity, but the order is roughly the same. For polymers, one has to take also into account their solubility in the corresponding solvent or solvent mixture. In Table 3.2, some typical HPLC solvents are listed:

It is obvious that the polarity may vary considerably within each group even for chemically quite similar solvents. Another limitation concerns detection. If a UV detector is used, one has to take into account the absorption of the solvent (UV-cutoff): as a rule of thumb, it should not exceed 1.0 A.

Table 3.2. Typical solvents used in liquid chromatography of polymers

Class	Solvent	Polarity index	UV cutoff
Alkanes	Hexane,Heptane	0,0	200
	Cyclohexane	0,2	200
Aromatics	Benzene	2,7	280
	Toluene	2,4	285
	Xylene	2,5	290
Ether	Diisopropyl ether	2,2	220
	Methyl-tert.-butyl ether	2,5	210
	Tetrahydrofuran	4,0	215
	Dioxane	4,8	215
Alkyl halides	Tetrachloromethane	1,6	263
	Dichloromethane	2,5	235
	Dichloroethane	4,0	225
	Trichlormethane	4,8	245
Esters	Butyl acetate	4,0	254
	Ethyl acetate	4,4	260
Ketones	Methyl ethyl ketone	4,7	329
	Acetone	5,1	330
Alcohols	n-Butanol	3,9	215
	i-Propanol	3,9	210
	n-Propanol	4,0	210
	Methanol	5,1	205
Nitriles	Acetonitrile	5,8	190
Amides	Dimethylformamide	6,4	268
Carboxylic acids	Acetic acid	6,2	230
Water		9,0	200

3.5 Detectors

When the separated polymer molecules leave the column, they have to be detected by one or more detectors, the signal of which must represent the concentration of the polymer with good accuracy. In the analysis of polymers by SEC, only a limited number of the numerous HPLC detectors can reasonably be applied. Basically, one has to distinguish between the following groups of detectors:

Concentration sensitive detectors		Molar mass sensitive detectors
Selective detectors	**Universal detectors**	
UV detector	RI detector	Single capillary viscometer
IR detector	Conductivity detector	Differential viscometer
Fluorescence detector	Density detector	Light scattering detectors:
Electrochemical detector	Evaporative light scattering detector	LALLS, MALLS, MALLS3, TALLS, RALLS

3.5.1 Concentration-sensitive Detectors

The signal (response) of a concentration sensitive detector is determined by the concentration of the solute in the mobile phase. Among the concentration sensitive detectors, one has to distinguish between detectors measuring a property of the solute and detectors measuring a (bulk) property of the mobile phase. The first group can thus be regarded as selective, the second one as universal (even though this is not a general rule). In the analysis of copolymers, two detectors may be required to monitor the concentrations of the two comonomers.

In HPLC, various *selective detectors* are usual, but not all of them can be applied to polymers:

- Photometric detectors: UV, (FT)IR
- Fluorescence detector
- Electrochemical detector.

Among the photometric detectors, the UV-detector is the most frequently used instrument. IR detectors are very useful, but limited to certain mobile phases, which do not absorb at the detection wavelength. A good alternative is off-line coupling of FTIR with HPLC using an evaporative interface [2–4], which provides valuable qualitative information. In such a system, the eluate is sprayed onto a Germanium disc, which can be transferred to any FTIR spectrometer to yield the full spectral information over any peak of the chromatogram. It must, however, be mentioned that such a device should be combined with an additional concentration detector, because quantitation may be somewhat problematic otherwise. Fluorescence detection cannot be applied to most polymers, and the same is true for electrochemical detection. In the case of oligomers, these detectors may be used after derivatization of the end groups [5–7], which is, however, typically not feasible in chromatography of high polymers.

UV-Detector. The most familiar solute property detector is the UV absorption detector, which exists in different modifications and is available from most producers of HPLC instruments. This detector measures the absorption of light of a selected wavelength and can be applied to polymers containing chromophoric groups. Typical detection wavelengths are in the range of 180–350 nm, which can, however, be utilized only in solvents with a sufficiently low absorbance. Many typical SEC solvents (aromatics, esters, ketones, DMF, $CHCl_3$ etc.) do not allow UV detection at low wavelengths (<250 nm), and also in ethers the UV cutoff may be dramatically influenced by contaminations, which are formed by oxidation, such as peroxides in THF etc., or by stabilizers.

Basically, one has to distinguish between three types of UV-detectors, which differ in the way monochromatic light is obtained:

- Fixed-wavelength detectors
- Variable wavelength detectors
- Diode-array detectors.

In most fixed wavelength detectors, a low-pressure mercury lamp emitting at 254 nm wavelength is used as the light source. In variable wavelength detectors, monochromatic light is obtained by means of a holographic grating. In some instruments, wavelength programming is also provided. Diode-array detectors allow simultaneous measurement of an entire UV-spectrum at any point of the chromatogram. The main difference between classical variable wavelength detectors and the diode-array detector (DAD) is the arrangement of monochromator and sample cell: in the DAD, the monochromator is placed in the light beam behind the sample cell ("inverse optics").

Basically, there are four types of *universal detectors*, which can be applied to HPLC:

- Refractive index (RI-) detector
- Conductivity detector
- Density detector
- Evaporative detectors.

Even though universal detectors are in general less sensitive than selective detectors, they are applied in the analysis of many polymer samples, such as polyolefins, aliphatic polyethers or the like, which cannot be detected by a selective detector. The most familiar instrument in this group is the RI detector, which exists in various modifications. The conductivity detector is not very useful in SEC of polymers, at least not in non-aqueous SEC of polymers. The density detector (operating according to the mechanical oscillator principle) is very useful in polymer analysis, especially in combination with other detectors. Evaporative detectors vaporize the mobile phase, and the non-volatile components of the sample can be detected by measuring the scattering of a transversal light beam, as is the case in the evaporative light scattering detector (ELSD). It is also possible to use such an evaporation device as an interface to a mass spectrometer or an FTIR spectrometer [2–4]. In liquid chromatography of polymers, only the ELSD is used as a routine instrument.

RI Detector. There are basically three types of RI detectors, the working principle of which has been described in refs. [8, 9]:

- Fresnel refractometer
- Deflection refractometer
- Interferometric refractometer.

Deflection refractometers have a better linear range than Fresnel refractometers, but have larger measuring cells. Interferometric refractometers are more sensitive (by one order of magnitude) than the other RI detectors. It must be mentioned that response factors of the RI detector depend on molar mass and on chemical composition, as is also the case for the density detector. While molar mass dependence can be compensated rather easily (see Section 4.6), the use of a second concentration detector is inevitable in the analysis of copolymers or polymer blends, as will be discussed in Chapter 4. Moreover, preferential solvation of the polymer coils by one component of the mobile phase may affect detector signals [10, 11], which can also require a second concentration detector.

Density Detector. The density detector provides additional information in SEC, when combined with a UV or RI detector. This instrument utilizes the mechanical oscillator principle [12–14]. Its measuring cell is an oscillating, U-shaped capillary, the period of which depends on the density of its content. Period measurement is performed by counting the periods of an oven-controlled 10 MHz quartz during a predetermined number of periods of the measuring cell. The signal of such a detector is thus inherently digital, which means that no A/D converter is required in data acquisition, and its response is integrated over each measuring interval.

Evaporative Light Scattering Detector. Although the ELSD measures a property of the solute, it can be regarded as universal detector, because it detects any non-volatile components of a sample [15–17]. While there is a lot of UV-detectors on the market, only a few producers offer this type of detector. In such an instrument, the eluate is nebulized and the solvent evaporated from the droplets. Each droplet containing non-volatile material will form a particle. When the aerosol thus obtained crosses a light beam, the light will be scattered. This effect can be utilized for detection, even though there are still some problems in quantification of the signal. The working principle of such an instrument is illustrated by Fig. 3.5.

Even though such an instrument is very useful in LAC, because it allows gradient elution (provided that no non-volatile buffers have to be used), there are still some questions concerning quantitation. First of all, the linearity of the ELSD is sometimes poor (its response can, however, be expressed by an exponential func-

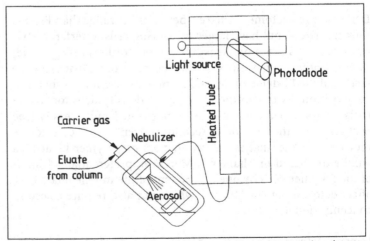

Fig. 3.5. Schematic representation of the evaporative light scattering detector (ELSD)

tion [18]). Moreover, the dependence of its sensitivity on molar mass and chemical composition of the sample is not yet clear. This is easy to understand, as the way from concentration to a detector signal is much more complex for this instrument than for other detectors: As the intensity of the scattered light will depend on the number, size, and refractive index of the particles, it must be influenced by several parameters concerning the sample, such as molar mass and chemical composition. Moreover, the size of the droplets formed in the nebulizer will depend on operation conditions and again on the chemical nature of the sample.

Hence some additional parameters influence the sensitivity, such as oven temperature, flow rate of carrier gas and eluate, viscosity and surface tension of the latter (which will of course change when a surface active substance is eluted, and also in gradient elution) etc. This means that an ELSD must be calibrated very carefully in order to yield reliable results, as has been shown by several authors [19–21].

3.5.2 Molar Mass Sensitive Detectors

Molar mass sensitive detectors are useful in SEC, because they yield the molar mass of each fraction of a polymer peak. Since the response of such a detector depends on both concentration as well as the molar mass of the fraction, it has to be combined with a concentration-sensitive detector. The following types of molar mass sensitive detectors are available [22–25]:

– Differential viscosimeters
– Low-angle light scattering detectors (LALLS)
– Two-angle light scattering detectors (TALLS)
– Right-angle light scattering detectors (RALLS)
– Multi-angle light scattering detectors (MALLS, MALLS3).

Since viscosity and light scattering yield different information, it makes sense to combine both of them [26]. From light scattering detection, the absolute molar mass distribution (MMD) can be determined directly. Using more than one angle, one may also obtain the radius of gyration. On the other hand, SEC with viscosity detection yields the intrinsic viscosity distribution (IVD). In this case, the MMD is, however, determined indirectly and is thus subject to retention errors. Hence, a combination of a concentration detector with both a light scattering detector and a viscosity detector provides the highest reliability of results. Moreover, information on branching can be obtained in this way [27, 28].

Viscosity Detector [29–34]. As has already been explained in Section 1.4, the viscosity of a polymer solution is closely related to the molar mass (and architecture) of the polymer molecules. Viscosity measurement in SEC can be performed by measuring the pressure drop P across a capillary, which is proportional to the viscosity η of the flowing liquid (the viscosity of the pure mobile phase is denoted as η_0). The relevant parameter is, however, the intrinsic viscosity $[\eta]$, which is defined as the limiting value of the ratio of specific viscosity ($\eta_{sp}=(\eta - \eta_0)/\eta_0$) and concentration c for c→ 0:

$$[\eta] = \lim_{c \to 0} \frac{\eta - \eta_0}{\eta_0 \cdot c} = \lim_{c \to 0} \frac{\eta_{sp}}{c} \qquad (3.1)$$

When a polymer passes the capillary, the pressure drop is increased by ΔP. In viscosity detection, one has to determine as well the viscosity η of the sample solution as also the viscosity η_0 of the pure mobile phase. The specific viscosity $\eta_{sp}=\Delta\eta/\eta$ is obtained from $\Delta P/P$. As the concentrations in SEC are typically very low, $[\eta]$ can be approximated by η_{sp}/c.

The simple approach using one capillary and one differential pressure transducer will not work very well, because the viscosity changes $\Delta\eta=\eta-\eta_0$will typically be very small compared to η_0, which means that one has to measure a very small change of a large signal. Moreover, flow-rate fluctuations due to pulsations of a reciprocating pump will lead to much greater pressure differences than the change in viscosity due to the eluted polymer. Instruments of this type should be used with a positive displacement pump.

A better approach is the use of two capillaries (C1 and C2) in series, each of which is connected to a differential pressure transducer (DP1 and DP2), and a sufficiently large holdup reservoir (H) in between. With this approach, one measures the sample viscosity η from the pressure drop across the first capillary, and the solvent viscosity η_0 from the pressure drop across the second capillary. Pulsations are eliminated in this setup, because they appear in both transducers simultaneously. Another design is the differential viscometer, in which four capillaries are arranged similar to a Wheatstone bridge. This detector measures the pressure difference ΔP at the differential pressure transducer between the inlets of the sample capillary and the reference capillary, which have a common outlet, and the overall pressure P at the inlet of the bridge. In Fig. 3.6, both designs are shown schematically.

In the "bridge" design, a holdup reservoir in front of the reference capillary (C4) makes sure that only pure mobile phase flows

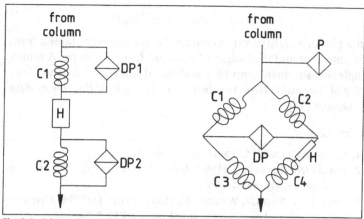

Fig. 3.6. Schematic representation of differential viscometers

through the reference capillary, when the peak passes the sample capillary (C3). This design offers considerable advantages: The detector measures actually the pressure difference ΔP at the differential pressure transducer (DP) between the inlets of the sample capillary and the reference capillary, which have a common outlet, and the overall pressure P at the inlet of the bridge. The specific viscosity $\eta_{sp}=\Delta\eta/\eta_0$ is thus obtained from $\Delta P/P$. One concern with this type of detector is that the flow must be divided 1:1 between both arms of the bridge. This is achieved by capillaries 1 and 2, which must have a sufficiently high back pressure. Nevertheless, when a peak passes through the sample capillary, a slight deviation of the 1:1 ratio will be observed. A problem of flow rate variations exists also in a single capillary viscometer: When the polymer peak passes through the measuring capillary, the increased back pressure leads to a peak shift (Lesec effect) [35].

Light Scattering Detector. In a light scattering detector, the scattered light of a laser beam passing through the cell is measured at angles different from zero. The (excess) intensity $R(\theta)$ of the scattered light at the angle θ is related to the weight-average of molar mass M_w:

$$\frac{K^*c}{R(\theta)} = \frac{1}{M_w P(\theta)} + 2A_2c \tag{3.2}$$

wherein c is the concentration of the polymer, A_2 is the second virial coefficient, and $P(\theta)$ describes the scattered light angular dependence. K^* is an optical constant containing Avogadro's number N_A, the wavelength λ_0, the refractive index n of the solvent, and the refractive index increment dn/dc of the sample:

$$K^* = 4\pi^2 (dn/dc)^2 /(\lambda^4 N_A) \qquad (3.3)$$

In a plot of $K^*c/R(\theta)$ versus $\sin2(\theta/2)$, M_w can be obtained from the intercept and the radius of gyration from the slope. A multi-angle measurement provides additional information. The accuracy of the results depends, however, strongly on the way of data treatment [36].

References

1. SNYDER LR (1978) J Chromatogr Sci 16: 223
2. WILLIS JN, WHEELER L (1993) Abstracts of papers of the American Chemical Society, PMSE 206: 61
3. PROVDER T, KUO CY, WHITED M, HUDDLESTON D (1994) Abstracts of papers of the American Chemical Society, PMSE 208: 186
4. SCHUNK TC, BALKE ST, CHEUNG P (1994) J Chromatogr A: 661: 227
5. NITSCHKE L, HUBER L (1993) Fresenius J Anal Chem 345: 585
6. KIEWIET AT, VANDERSTEEN JMD, PARSONS JR (1995) Anal Chem 67: 4409
7. DESBENE PL, DESMAZIERES B (1994) J Chromatogr A661: 207
8. YAU WW, KIRKLAND JJ, BLY DD (1979) Modern Size Exclusion Chromatography. Wiley, New York, p 148
9. BELENKII BG, VILENCHIK LZ (1983) Modern Liquid Chromatography of Macromolecules. J Chromatogr Libr 25: 199
10. TRATHNIGG B, YAN X (1993) J Chromatogr A 653: 199
11. TRATHNIGG B, THAMER D, YAN X, MAIER B, HOLZBAUER H-R, MUCH H (1994) J Chromatogr A 665: 47
12. TRATHNIGG B, JORDE C (1987) J Chromatogr 385:17
13. TRATHNIGG B (1990) J Liq Chromatogr 13(9):1731
14. TRATHNIGG B (1991) J Chromatogr 552: 507
15. LAFOSSE M, ELFAKIR L, MORIN-ALLORY L, DREUX M (1992) J High Res Chromatogr 15: 312
16. RISSLER K, FUCHSLUEGER U, GRETHER HJ (1994) J Liq Chromatogr 17: 3109
17. BROSSARD S, LAFOSSE M, DREUX M (1992) J Chromatogr 591: 149
18. DREUX M, LAFOSSE M, MORIN-ALLORY L (1996) LC-GC Int 9: 148
19. VANDERMEEREN P, VANDERMEEREN J, BAERT L (1992) Anal Chem 64: 1056
20. HOPIA AI, OLLILAINEN VM (1993) J Liq Chromatogr 16: 2469
21. TRATHNIGG B, KOLLROSER MJ (1997) J Chromatogr., 768: 223
22. PODZIMEK S (1994) J Appl Polym Sci 54: 91
23. DAYAL U, MEHTA SK (1994) J Liq Chromatogr 17: 303
24. DAYAL D (1994) J Appl Polym Sci 53: 1557
25. BALKE S, RAO B, THITIRATSAKUL R, MOUREY TH, SCHUNK TC (1996) Prep 9th Int Symp Polym Anal Char (ISPAC 9), Oxford, July 1-3
26. JACKSON C, YAU WW (1993) J Chromatogr 645: 209

27. JACKSON J (1994) J Chromatogr A 662: 1

28. PANG S, RUDIN A (1993) Chromatography of polymers ACS Symp. Ser. 521, American Chemical Society, New York, p 254

29. YAU WW, ABBOTT SD, SMITH GA, KEATING MY (1987) In: Provder T (ed) Detection and Data Analysis in Size Exclusion Chromatography. ACS Symp Ser 352, American Chemical Society, Washington, DC, p 80

30. STYRING MG, ARMONAS JE, HAMIELEC AE (1987) In: Provder T (ed) Detection and Data Analysis in Size Exclusion Chromatography. ACS Symp Ser 352, American Chemical Society, Washington, DC, p104

31. HANEY MA, ARMONAS JE, ROSEN L (1987) In: Provder T (ed) Detection and Data Analysis in Size Exclusion Chromatography. ACS Symp Ser 352, American Chemical Society, Washington, DC, p119

32. BROWER L, TROWBRIDGE D, KIM D, MUKHERJEE P, SEEGER R, McINTYRE D (1987) In: Provder T (ed) Detection and Data Analysis in Size Exclusion Chromatography, ACS Symp Ser 352, American Chemical Society, Washington, DC, p155

33. MORI S (1993) J Chromatogr 637: 129

34. CHEUNG P, LEW R, BALKE ST, MOUREY TH (1993) J Appl Polym Sci 47: 1701

35. LESEC J, MILLEQUANT M, HAVARD T (1993) In: Provder T (ed) Chromatography of Polymers. ACS Symp Ser 521, American Chemical Society, New York, p 220

36. JENG L, BALKE ST, SNYDER LR(1993) J Appl Polym Sci 49: 1359

4 Size Exclusion Chromatography

SEC is a standard technique that is frequently used to determine MMD of polymers and the corresponding number, weight, and z averages, M_n, M_w, and M_z. As has already been mentioned in Section 2.2, the separation in SEC is governed by entropic effects, unlike other modes of liquid chromatography, where enthalpic effects are predominant. The separation mechanism is, however, not the only difference between these techniques: The main difference between SEC and other modes of LC is that the information about the samples is coded into the *absolute* peak position (while in LAC the relative elution order is sufficient) and the peak shape (which is generally very different from peaks in LAC). As molar masses and polydispersity of macromolecules are directly obtained from these chromatographic raw data, the requirements in data processing are considerably different from those in other chromatographic techniques. There are also consequences for instrumentation, maintenance, and operation that will be discussed in the following sections.

The principle of SEC is explained best by a schematic representation (Fig. 4.1), which shows the first part of an analysis by SEC: from the sample to a chromatogram.

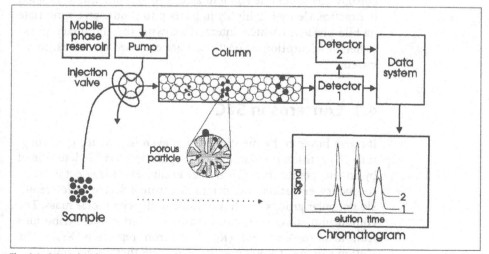

Fig. 4.1. Principle of size exclusion chromatography

When a sample is injected into the column, the polymer molecules are separated according to their hydrodynamic volumes (not actually to their molar mass): polymer coils larger than the largest pores of the packing cannot enter the pores; hence they are eluted at the interstitial volume V_i. No fraction of the sample can be eluted before the interstitial volume V_i has passed through the column. This elution volume corresponds to the exclusion limit of the column. Small molecules, which have access to the entire pore volume V_p, will appear at an elution volume equal to the sum of the interstitial volume V_i and the pore volume V_p. Molecules of a size between these extremes have access only to a part of the pore volume; hence they will be eluted at an elution volume V_e, which is given by Eq. 4.1:

$$V_e = V_i + K_{SEC} V_p \tag{4.1}$$

wherein K_{SEC} is the equilibrium constant of a sample in size exclusion chromatography ($0 < K_{SEC} < 1$). Since the SEC separation has to be performed within a volume considerably smaller than the total volume of the column, good separation efficiencies can only be achieved with rather long columns. The relation between K_{SEC} and the molar mass of a polymer is determined by a calibration, as will be discussed.

As described in Section 2.1, the equilibrium constant of a chromatographic separation can be correlated with thermodynamic parameters. The driving force for a separation at the (absolute) temperature T is the change in Gibbs free energy ΔG, which results from the changes in enthalpy and entropy, ΔH and ΔS, respectively. In ideal SEC, which should be governed only by entropy, ΔH should be equal to zero, which is not always fulfilled in practice. Hence it is highly important to choose an appropriate mobile phase, in which interactions with the stationary phase (whether adsorption or partition) are excluded or at least minimized.

4.1 Concerns in SEC

It must, however, be mentioned that even in absence of adsorption or partition phenomena the separation can be determined by an effect other than (ideal) size exclusion. This effect is called secondary exclusion and originates from (electrostatic) repulsion of polar groups and has nothing to do with molar mass. The equilibrium constant K can then be divided into contributions from ideal size exclusion (K_{SEC}) and from repulsion (K_{Rep}). The elution volume of a polymer molecule will then be given by

$$V_e = V_i + K_{SEC}K_{Rep} \cdot V_p \qquad (4.2)$$

If also interactions with the stationary phase occur (such as adsorption or partition), an additional term has to be taken into account, which consists of the equilibrium constant K_{LAC} of this interaction and the volume V_{stat} of the stationary phase:

$$V_e = V_i + V_p \, K_{SEC} \, K_{Rep} + V_{stat} \, K_{LAC} \qquad (4.3)$$

It is not always easy to distinguish between these effects.

Craven [1] has investigated the elution behaviour of polyoxyethylenes with different end groups (diols, mono- and dimethyl ethers) on a PL gel column in different mobile phases and found considerably different calibration lines for the different homologous series in different mobile phases. These differences were explained by combinations of exclusion with partition-adsorption effects. Mori and Nishimura [2] observed polyelectrolyte effects in SEC of poly(methyl methacrylate) (PMMA) and polyamides in hexafluoro-2-propanol, which could be suppressed by the addition of sodium trifluoroacetate as an electrolyte, which breaks down hydrogen bonding.

Another concern in SEC is the problem of peak dispersion. When a monodisperse sample is analysed by chromatography, it will appear as peak of a – more or less – Gaussian shape and not as a rectangular concentration profile (which it was immediately after injection). The main reasons for peak broadening are diffusion phenomena in the column, the capillaries, and the detector, which can be minimized, but not completely avoided. Asymmetric broadening can be caused by too high sample loads, interaction of the sample with the column packing, and an imperfect chromatographic system. Void volumes between the connecting capillaries will lead to a dramatically decreased performance of the system. While in other LC techniques peak spreading affects only resolution, it may have much more dramatic effects in SEC, where peak shape is more important than peak area (which is of more interest in other HPLC applications).

Basically, a chromatographic peak can be described by the function F(v), the detector response at a given elution volume. It must be mentioned that the actual concentration is not always easily obtained from F(v), as will be discussed later. This function results from a convolution of two other functions, $G_N(v,y)$, which is the shape function of a solute eluting at the mean elution volume y, and W(y), the chromatogram corrected for band spreading:

$$F(v) = \int_0^\infty W(y) G_N(v, y) dy \qquad (4.4)$$

This equation is well known in SEC as the Tung axial dispersion equation [3]. Various approaches of correcting chromatograms for peak dispersion have been published that work more or less well [4, 5]. It is clear that the deconvolution – the calculation of W(y) from F(y) and $G_N(v,y)$ – suffers from the problem that $G_N(v,y)$ is not easily obtained. However, the goal in SEC is to obtain low or negligible peak spreading, which is preferable to highly sophisticated algorithms for correcting for poor chromatograms. Moreover, other sources of error, such as too high sample loads, flow variations, improper baseline selection, uncorrected molar mass dependence of response factors, etc. will affect the accuracy of results much more. Correction of peak spreading makes sense only when all other sources of error have been taken into consideration.

As can clearly be seen from Eq. 4.1, the separation in ideal SEC is achieved within the pore volume of the column packing. Consequently, the separation is mainly determined by column length (unlike in LAC, where the composition of the mobile phase is the key parameter).

4.2 Equipment and Materials

Unlike other modes of HPLC, the SEC separation is determined exclusively by the stationary phase, while the mobile phase should have no effect – at least in the ideal case. The entire separation occurs within the volume of the pores, which typically equals only 30–40% of the total column volume. This means that a good separation efficiency requires much longer columns or often sets of several columns. While there is a strong trend towards microcolumns in LAC, the typical column lenghts in SEC are 25–60 cm, and often sets of 2 to 4 columns are used. Commercially available SEC columns usually have diameters of 5–8 mm for analytical separations and typically 22–25 mm for semipreparative applications. Recently, packings with smaller particles have become available that allow higher packing densities, thus providing higher plate numbers per meter of column. In Table 4.1, the average plate numbers per meter, as specified by different producers, and the corresponding plate heights are given for packings with different particle size.

Obviously, smaller particles will provide a much better separation efficiency. On the other hand, packings with smaller particles build up a higher back pressure.

Table 4.1. Average plate numbers per meter and corresponding plate heights for PS-DVB packings from different producers [6]

Particle size (µm)	Plate number N per m	Plate height H (mm)
3	>80000	0.0125
5	>45000	0.0222
10	>30000	0.0333
20	>15000	0.0667

Besides the packing density, back pressure depends on the viscosity of the mobile phase and its linear velocity, which is inversely proportional to the column diameter and directly proportional to the (volume) flow rate. For most packings, the back pressure should not exceed 150 bar, because that would reduce their life time or destroy them immediately.

It is a well known fact that there is an optimum flow rate (or better: linear velocity) of the mobile phase, at which the highest plate number N (or the lowest plate height) is achieved. The height H of a theoretical plate is given by the Van Deemter equation

$$H = A + \frac{B}{v} + C \cdot v \qquad (4.5)$$

wherein v is the flow velocity and A,B and C are constants associated with the terms due to eddy diffusion, longitudinal diffusion, and mass transfer [7]. Eddy diffusion does not depend on v, longitudinal diffusion decreases (hyperbolically) with inreasing v, and mass transfer increases linearly with v; hence the summation of these three terms results in a curve with a minimum.

As a rule of thumb, the optimum (volume) flow rate for a column of 8 mm diameter is about 1 mL/min, for a 4 mm column 0.25 mL/min and so on. Consequently there are some limitations in the choice of column dimensions:

- In a given mobile phase, the overall length is limited by the back pressure produced at a flow rate close to the optimum.
- Further reduction of the (volume) flow rate is not a good solution, because on this side of the Van Deemter equation the plate height increases very rapidly.
- In general, it is better to apply higher temperatures, which reduce the viscosity of the mobile phase and thus the back pressure.

Stationary Phases. The most frequently used column packings are either based on porous silica or on semirigid (highly crosslinked) organic gels, in most cases copolymers of styrene (St) and divinylbenzene (DVB). These packings are suitable for non-aqueous organic solvents. In aqueous SEC, modified silica or crosslinked hydrophilic polymers may be used. There are, however, also polymer-based packings available, which can be used in different mobile phases. In general, silica-based packings are rather rugged, while organic packings have to be handled very carefully, as will be pointed out later. In Tables 4.2 and 4.3, typical column packings for both aqueous and non-aqueous SEC and their producers are given.

Typically the separation range of a column covers approximately two decades of molar mass. Most producers also offer mixed-bed columns, which cover a much wider range (e.g. from 1×10^2 to 1×10^6). These columns provide typically a better linear calibration than combinations of single columns.

Mobile Phases. Mobile phases in SEC are normally single solvents and should meet the following requirements:

– thermodynamically good solvent for the polymer
– solvate (wet) the packing
– chemically inert towards the chromatographic system and sample
– easily purified and redistilled
– low viscosity
– good UV transparency, if needed
– appropriate refractive index (with respect to sample)
– low toxicity.

Obviously, the mobile phase must be a good solvent for the polymer, because otherwise non-exclusion effects would severely affect the separation. Depending on the nature of the polymer samples (and the columns), water or organic solvents can be used (see also Section 3.4, Table 3.2). The most frequently used organic solvents are THF, chloroform, toluene, esters, ketones, DMF etc. In the analysis of rather polar polymers, such as polyelectrolytes, one may add electrolytes in order to reduce non-exclusion effects. For many polymers, such as polyolefins, PET, etc., there are only a few good solvents, and often one has to work at higher temperatures.

Table 4.2. Column packings for non-aqueous SEC

Producer	Packing	Material
Jordi	Jordi GPC	100% DVB polymer
Macherey-Nagel	Nucleogel GPC	St-DVB copolymer
Merck	LiChrogel PS	St-DVB copolymer
Phenomenex	Phenogel	St-DVB copolymer
Polymer Laboratories	PL gel	St-DVB copolymer
Polymer Standards Service	SDV	St-DVB copolymer
	PFG (Polar Fluoro Gel)	surface modified silica
Rockland Technologies	Zorbax PSM	porous silica microspheres, silanized
Shodex	Asahipak GF HQ	highly crosslinked polyvinyl alcohol
Shodex	GPC K, HT, UT	not available from manufacturer
Waters	Styragel HR, HT, MW, Ultrastyragel, μ-Styragel	St-DVB copolymer

Table 4.3. Column packings for aqueous SEC

Producer	Packing	Material
Macherey-Nagel	Nucleogel	silica based
Merck	LiChroSpher Diol	silica, diol-modified
Phenomenex	W-Porex GP	silica based
Polymer Laboratories	PL aquagel OH	not available from manufacturer
Polymer Standards Service	HEMA Gel HEMA Bio Gel	hydroxyethyl methacrylate copolymers
Rockland Technologies	Zorbax PSM	porous silica microspheres, unsilanized
Shodex	OH-Pak SB 800 series	crosslinked poly-(hydroxyethyl-methacrylate)
Shodex	Protein KW 800 series	silica
Shodex	Chitopak KQ 802.5	chitosan
Shodex	Asahipak GF HQ OH-Pak Q 800 series	highly crosslinked polyvinyl alcohol
TosoHaas	TSK PW	crosslinked polymer
TosoHaas	TSK SW	modified silica
Waters	Ultrahydrogel	not available from manufacturer

4.3 Data Acquisition and Processing

The basic steps in processing SEC data can be described best by a graphic representation, which shows the way from a chromatogram to the MMD. The molar mass for each point of the peak can be obtained from a calibration function or a molar mass sensitive detector. The MMD of the sample can be calculated from the chromatographic raw data (Fig. 4.2).

However, there are numerous sources of error and pitfalls in the separation itself, in the detection, and also in the treatment of the chromatographic raw data. Provided that the separation itself is reliable, the subsequent transformations may be subject to errors:

1. **Elution time to elution volume conversion:** As only slight variations of flow rate may cause severe errors in molar mass, a highly constant and reproducible flow rate must be maintained during the entire chromatogram. This requires a high quality pump. Changes in flow rate between chromatograms can be compensated by using a (low molar mass) internal standard as flow rate marker.

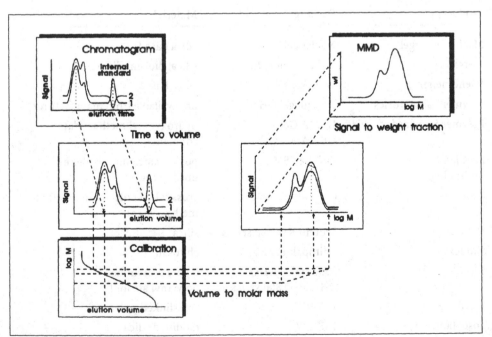

Fig. 4.2. Basic steps in obtaining a molar mass distribution from a chromatogram

2. **Elution volume to molar mass conversion:** The molar mass of a fraction is typically obtained from a calibration, the quality of which determines the reliability of the results. Alternatively, molar mass sensitive detectors can be used (in addition to the concentration detector).

3. **Detector response to polymer concentration conversion:** This requires a sufficiently wide linear range, a well-defined response of the detector(s) along the entire peak: in copolymers or even in low molar mass homopolymers response factors may vary within the peak. For homopolymers, molar mass dependence of detector response may be corrected rather easily, while in the case of copolymers typically a second concentration detector is required.

4. **Polymer concentration to weight fraction:** The final goal of such an analysis is the determination of the molar mass distribution (MMD), which can be correlated with physical properties. This simple normalization procedure can, however, make problems at the ends of the peak, if the baseline and the integration limits are not correct.

In the following sections, each of these steps will be referred to in detail.

4.3.1 Operation of an SEC System

In this section, requirements concerning sample treatment, the chromatographic equipment, data acquisition and processing will be discussed and the different approaches for the analysis of different polymers evaluated.

The Pump. As has already been pointed out in Section 3.1, different types of pumps may be used in SEC, the performance of which may be different. In general, positive displacement pumps should be the first choice in SEC, especially in systems with a viscosity detector. However, reciprocating pumps are commonly used in SEC and perform quite well. With a pulsation dampener, a sufficiently smooth flow can be achieved.

Column Selection. When selecting columns for a given separation problem in SEC, one may choose from a large number of columns from different producers. In general, the following considerations should be taken into account.

– The separation range should be selected carefully: it does not make sense to use a column with an exclusion limit of 10^6 for the analysis of oligomers. For high molar mass polymers, the high molecular end of the MMD should be within the separation range.

– The particle size has to be taken into account, since it determines the plate height. Small particles (typically 5 µm) provide higher plate numbers and thus the same separation efficiency with a lower overall column length than columns packed with larger particles (10 µm), but produce a higher back pressure for a given column length. Shorter columns save time and solvent. However, the life time of 5 µm (or even 3µm) packings may be considerably reduced by samples containing contaminants that may deposit in or remain on the column.

– Combinations of packings with a different separation range can be achieved either by using columns with different porosity or by mixed-bed columns, which typically provide a better linear calibration than combinations of columns.

When combining columns to a set, it is better to combine fewer longer than more shorter columns, because the column ends as well as the connecting capillaries increase peak broadening.

– Some packings must not be used in certain mobile phases or at higher temperatures, which are required for polymers of low solubility, such as polyolefins.

The Mobile Phase. The mobile phase in SEC must be a good solvent for the polymer in order to avoid non-exclusion effects, as has already been pointed out. It is also important to dissolve the sample sufficiently long before injecting it in order to allow the polymer coils to swell in the solvent. To avoid air bubbles in the system, the mobile phase should be degassed in vacuo, by ultrasound, or helium. The use of a degasser, which is available from several distributors, may also be advantageous.

Even though SEC is typically performed in one-component mobile phases, one has to keep in mind that the purity of even HPLC grade solvents is typically 99.5–99.9 %. Moreover, some solvents are hygroscopic, form peroxides (such as THF) or contain a stabilizer (like chloroform, which is stabilized by up to 1% ethanol or 2-methylbutene). Low amounts of a second component in the mobile phase can lead to integration errors, which are due to preferential solvation of the polymer molecules. This effect has been known for a long time [8, 9], but it is often neglected. In a mixed solvent, a polymer coil may take up one component of the solvent selectively. When the sample is separated on the column, a system peak (vacancy peak) appears. Obviously, the missing amount of solvent in the system peak appears in the peak of the polymer, the area of which is different from what it would have been in the absence of preferential solvation.

Preferential solvation may depend on molar mass, which means that it may affect the accuracy of the MMD [10]. Hence it is important to dissolve the sample in the solvent from the SEC reservoir. If a solvent peak is observed, this is a strong hint for preferential solvation.

Handling SEC Columns. SEC columns are handled as follows: A column set in SEC should always be used in the same mobile phase, because solvent changes can reduce column life and performance. If, however, a solvent change is necessary, this should be done stepwise and at a low flow rate (0.5 ml/min maximum). For some solvents, a direct change should be avoided by using an intermediate solvent. It should be mentioned that changing the mobile phase always requires a recalibration.

- SEC columns should not be operated in the backwards direction, because this can damage the column packing.
- Care should also be taken in connecting columns or in sample injection.
- Replacing a clogged inlet frit can also reduce column performance considerably. When analysing samples, which may contaminate the column, one should always use a precolumn.
- Pulsations from the pump, which can be caused by air bubbles in the solvent line, a leakage of one pump seal, or a damaged or dirty check valve, can also reduce column life and performance.

Determination of Elution Volumes. As mentioned above, three transformations have to be performed with the chromatographic raw data. The first one – time to volume – is performed by a multiplication of elution times with the flow rate. Hence a highly constant flow rate has to be maintained during the entire chromatogram. This is very important in SEC: due to the logarithmic relation between molar mass and elution volume a change of the flow rate of only 0.1% can cause an error in molar mass of up to 10%.

The flow rate precision of most pumps is, however, typically 0.2–0.3%, and this precision can be further reduced by leakages in the system, air bubbles in the solvent line, dirty or damaged check valves, or increased back pressure from the column. An inline filter in the solvent reservoir can prevent particles from coming into the pump heads. However, even stainless steel filters may corrode in some solvents (such as chloroform). Obviously, rust particles may damage check valves and pump seals. The absolute flow rate (in the calibration) can be obtained by measuring the time to fill a calibrated flask or by weighing the mobile phase eluting in a defined time.

Table 4.4. Errors in molar mass averages due to flow rate deviations

Flow rate (mL/min)	Deviation (%)	M_w (exp)	ΔM_w (%)	V(int.St.)(mL)
1.020	+2.00	48200	+36.1	21.80
1.013	+1.30	43400	+23.6	21.65
1.006	+0.60	39300	+11.0	21.50
1.000	0	35400	0	21.37
0.992	−0.80	31100	−12.1	21.20
0.985	−1.50	27700	−21.8	21.05
0.978	−2.20	24300	−31.4	20.90

There have been attempts to determine the flow rate by measuring the time of a thermal pulse along a capillary, but the precision of these devices is not sufficient. The more efficient – and less expensive – approach is the use of a low-molar mass internal standard in the MMD-calibration and in each chromatogram. The corrected flow rate is obtained from the ratio of the elution times of this standard peak. In a systematic study, Kilz has demonstrated the effect of flow rate changes and the use of the internal standard method. A polystyrene standard was analysed on a typical column set (2 columns 8x300 mm, PSS SDV 5μm, 1000 Å + 10^5 Å) in THF at different flow rates (measuring temperature: 25 °C, sample concentration: 1 g/L, 20 μL sample volume). The results are given in Table 4.4 [11].

Obviously, the error in M_w is very large even at flow rate deviations below 1%, and this can even be worse with linear columns, the calibration function of which typically has a higher slope. Using the elution volume of the internal standard peak, one may easily calculate the actual flow rate, thus avoiding these errors.

Determination of Molar Mass. The second transformation – volume to molar mass – requires either a calibration or the use of a molar mass sensitive detector. Unless a molar mass sensitive detector is used, one has to determine the molar mass of a fraction eluting at volume V_e by calibration, which is discussed in the following section.

4.3.2 Calibration

Calibration with Narrow Standards. The easiest way to establish a calibration, from which the molar mass for a given elution volume can be obtained, is the peak position method, which can be applied if a series of standards with a narrow MMD is available. Since appropriate standards have become commercially available for many polymers, calibration with narrow standards can be applied to many types of polymers. Some suppliers provide well-characterized standards for speciality polymers, also. In Table 4.5, a list of available standards and their suppliers is given.

In the low molar mass range, additional data points can be taken from the maxima of oligomer peaks, which are at least partially resolved. If one of these peaks can be identified, the peaks of the higher oligomers may also be used in the calibration (see Fig. 4.16). An extension to even higher molar masses can be achieved by semipreparative separation of oligomers by liquid adsorption chromatography [12].

In the classical approach, one assumes a linear relation between the logarithmic molar mass and elution volume. The calibration function is

$$\log M = A + B \cdot V_e \qquad (4.6)$$

wherein A and B are constants that can be determined by linear regression.

The assumption of a linear relation between $\log M$ and V_e is, however, only a first approximation, which depends on the columns used. For many columns, the calibration lines are rather sigmoidal than linear. In this case, a polynomial fit can match the experimental points much better [13].

$$\log M = A + B V_e + C \cdot V_e^2 + D \cdot V_e^3 + E \cdot V_e^4 \ldots \qquad (4.7)$$

The coefficients A-E in such a relation can be determined by linear regression (even though this equation is non-linear in V), a feature that should be provided by commercially available software packages. The order of the polynomial fit may, however, be critical: in some cases, a fit of high order may produce an erroneous calibration line. In general, the standards should be roughly equidistant on the volume scale and cover a sufficiently wide molar mass range. The software should allow an inspection of the fit. In the analysis of samples, for which narrow MMD standards are not available, the following approaches can be used.

Calibration with Broad Standards. If no standards with a sufficiently narrow MMD are available, a calibration can also be obtained with broad MMD standards. Depending on the available reference materials different strategies for establishing a calibration can be applied [14, 15].

Table 4.5. Narrow MMD standards for SEC calibration and suppliers:(a) Polymer Laboratories, (b) Polymer Standards Service, (c) Waters Corp.

Polymer	Supplier	number of standards	lowest	highest
Polystyrene	a	29	162	15,000,000
Polystyrene	b	18	500	4,000,000
Polystyrene	c	20	400	20,000,000
Poly(methyl styrene)	b	6	10,000	700,000
Poly (2-vinyl pyridine)	b	9	2,900	1,000,000
Poly (methyl methyacrylate)	a	20	500	1,500,000
Poly (methyl methyacrylate)	b	>70	102	1,200,000
Poly (methyl methyacrylate)	c	10	1,000	15,000,000
Poly(ethyl methyacrylate)	b	10	2,000	500,000
Poly (n-butyl methyacrylate)	b	10	1,000	750,000
Poly (t-butyl methyacrylate)	b	20	1,000	1,000,000
Poly(t-butyl vinyl ketone)	b	5	10,000	450,000
Polyisoprene	a	11	1,000	3,000,000
Polyisoprene	c	10	1,000	3,000,000
Poly(isoprene-1,4)	b	14	700	800,000
Poly(isoprene-3,4)	b	4	900	250,000
Polybutadiene	a	11	1,000	1,000,000
Polybutadiene	b	4	5,000	80,000
Polybutadiene	c	10	1,000	1,000,000
Polyethylene	a	10	170	120,000
Polyvinylchloride	b	10	100	320,000
Polyisobutylene	b	16	112	1,000,000
Oligoethylene, linear	b	8	112	2,200
Poly(ethylene)	b	10	100	170,000
Poly(dimethylsiloxane)	b	11	162	205,000
Polycarbonate	b	22	300	200,000
Poly(tetrahydrofuran)	a	10	1,000	500,000
Poly(tetrahydrofuran)	c	10	1,000	1,000,000
Poly(propylene oxide)	c	3	1,200	4,000
Poly(ethylene oxide)	a	21	106	1,500,000
Poly(ethylene oxide)	b	18	106	1,000,000
Poly(ethylene oxide)	c	17	106	1,300,000
Poly(methacrylic acid), Na-Salt	b	20	1,000	800,000
Poly(acrylic acid), Na-salt	b	12	1,000	3,000,000
Polystyrene-sulfonate, Na-salt	b	15	200	3,000,000
Dextrane	b	10	180	300,000

The Integral-MMD method can be applied if the entire MMD of the standard is known with high accuracy (which is, however, seldom the case). The method compares – point per point – the cumulative MMD $F_N(v)$ of the standard and the cumulative distribution form of the chromatogram, $F_{N,cum}(v)$, which is calculated from

$$F_{n,cum}(v) = \int_0^v F_N(v)dv \qquad (4.8)$$

No assumptions on the shape of the calibration are made, the value of the method depends, however, strongly upon how the MMD of the standard was determined. If the MMD is very accurately known, the main source of error is the effect of band spreading on the chromatogram heights, which are problematic especially at the ends of the MMD of the standard.

If only the molar mass averages of one or more standards are known from independent methods (light scattering or osmometry), linear calibration methods [16] can be applied. In its simple form, this approach assumes a linear calibration (Eq. 4.6), the parameters A and B are fitted to yield best agreement with M_w and M_n of the standard. It is clear that with just two known parameters only a linear calibration can be obtained, which is defined by two parameters (slope and intercept).

If more than one standard is available, an objective function with more than two parameters can be formulated, which is, however, not very easy: using the standards indvidually, one will obtain different calibration functions. The main problem in such an approach is the correlation between the parameters. McCrackin [17] has used the objective function

$$O(\theta_i) = \sum_{i=1}^m w_i \left[\left(\overline{M}_{q,EXP} - \overline{M}_q(\theta_i) \right) \right]^2 \qquad (4.9)$$

wherein M_q represents the qth molecular weight average (q=1: M_n, q=2: M_w etc.), EXP refers to the known value, θ_i are the searched parameters, and $w_i = 1/M_q^2$ the assigned values of the weighting factors. There have been many papers descibing different iterative methods for accomplishing the parameter estimation.

Calibration with broad standards yields either an *effective* calibration (to account for both calibration and band broadening correction) or the *true* calibration function, which would be found with narrow MMD standards. In the latter case, resolution correction has to be performed.

Universal Calibration. There can be considerable differences between the calibration lines for different polymers on the same column in the same mobile phase (see Fig. 4.6). This is especially important in the analysis of copolymers or polymer blends. Consequently, different molar masses will elute at the same volume, when a mixture of two homopolymers is analysed by SEC. The elution volume of a block copolymer should be between the elution volumes of the homopolymers of the same molar mass. If the composition of the copolymer at each point of the peak is known, a good approximation will be achieved by interpolation between the calibration lines.

A very elegant approach is based on the fact that in SEC the elution volume V_e of a polymer depends on its hydrodynamic volume, which is proportional to the product of its molar mass M and intrinsic viscosity $[\eta]$. In a plot of log $(M\,[\eta])$ versus V_e (obtained on the same column), identical calibration lines should be found for two polymers (1 and 2), which is considered as universal calibration [18].

$$M_1\,[\eta]_1 = M_2\,[\eta]_2 \tag{4.10}$$

The intrinsic viscosity is a function of molar mass, which is described by the Mark-Houwink relationship , wherein K and a are coefficients for a given polymer in a given solvent at a given temperature.

$$[\eta] = KM^a \tag{4.11}$$

Combination of Eqs. 4.10 and 4.11 yields

$$K_1 M_1^{a_1+1} = K_2 M_2^{a_2+1} \tag{4.12}$$

If a column has been calibrated with polymer 1 (e.g. polystyrene), the calibration line for another polymer (2) can be calculated, provided that the coefficients K and a are known for both polymers with sufficient accuracy:

$$\ln M_2 = \frac{1}{1+a_2}\ln\frac{K_1}{K_2} + \frac{1+a_1}{1+a_2}\ln M_1 \tag{4.13}$$

The concept of the universal calibration would provide an appropriate calibration also for polymers for which no narrow standards exist. For lower molecular weights the Dondos-Benoit relation should be used, which is linear in this range [19].

$$\frac{1}{[\eta]} = -A_2 + \frac{A_1}{\sqrt{M}} \tag{4.14}$$

The accuracy of K and a is, however, limited even in the case of polymers for which a sufficient number of well defined standards exists. There are very high variations in the values reported in the literature [20, 21]. Even for such common polymers as for polystyrene and poly(methyl methacrylate) the values may differ considerably.

The situation is even worse for unusual polymers: an accurate determination of K and a requires a sufficient number of narrow MMD standards. Where these standards are not available, the universal calibration does not bring about a considerable improvement. There have also been efforts to establish a conventional calibration with broad standards by searching for the intrinsic viscosity parameters [22]. After all, the expense for buying (even costly) narrow standards will be worth while in most cases. If such standards are not available, the method of choice will be the use of molar mass sensitive detectors.

4.3.3 Quantification in SEC

Once the first two transformations – time to volume and volume to molar mass – have been performed, there remains the third transformation – detector response to amount of polymer in a fraction – which can also be subject to errors, depending on the nature of the samples. In the following section, problems are referred to with respect to the type of polymer to be analysed: homo- or copolymer, oligomer or high molar mass polymer.

Molar Mass Dependence of Response Factors. As mentioned in Section 4.2, the detectors most frequently used in SEC are UV and RI detectors. The density detector is useful in the analysis of non-UV absorbing polymers. The UV-detector monitors UV-absorbing groups in the polymer, which may be the repeating unit, the end groups, or both. RI and density detectors measure a property of the entire eluate, that means, they are sensitive towards a specific property of the sample (the refractive index increment or the apparent specific volume, respectively).

It is well known [23] that specific properties are related to molar mass

$$x_i = x_\infty + \frac{K}{M_i} \qquad (4.15)$$

where x_i is the property of a polymer with molar mass M_i, x_∞ is the property of a polymer with infinite (or at least very high) molar mass, and K is a constant reflecting the influence of the end groups [24–26]. A similar relation holds for the response factors for RI and density detection.

$$f_i = f_\infty + \frac{K}{M_i} \qquad (4.16)$$

In a plot of the response factor f_i versus $1/M_i$ of a polymer homologous series (with the same end groups) a straight line is obtained with the intercept f_∞ (the response factor of a polymer with very high molar mass, or the response factor of the repeating unit) and the slope K, which represents the influence of the end groups [27, 28]. The magnitude of K determines the threshold in M, above which the molar mass dependence of response factors becomes negligible. Once f_∞ and K have been determined, the correct response factors for each fraction eluting from an SEC column can be calculated using this equation with the molecular weight obtained from the SEC calibration. In a recent paper [29], three methods have been described for the determination of f_∞ and K.

Copolymers and Polymer Blends. In the analysis of copolymers, the use of multiple detection is generally inevitable. If the response factors of the detectors for the components of the polymer are sufficiently different, the chemical composition of each slice of the polymer peak can be determined from the detector signals. Typically, a combination of UV and RI detection is used, but other detector combinations have been described also. If the components of the copolymer have different UV-spectra, a diode array detector will be the instrument of choice. One has, however, to keep in mind that non-linear detector response may also occur with UV-detection [30].

In the case of non-UV absorbing polymers, a combination of RI and density detection yields information on chemical composition [31–33]. The ELSD is not equally suitable because of its poor linearity [34] and its unclear response to copolymers. The principle of dual detection is rather simple: when a mass m_i of a copolymer, which contains the weight fractions w_A and w_B (=$1-w_A$) of the monomers A and B, is eluted in the slice i (with the volume ΔV) of the peak, the areas $x_{i,j}$ of the corresponding slice obtained from both detectors depend on the mass m_i(or the concentration $c_i = m_i/\Delta V$) of polymer in the slice, its composition (w_A), and the corresponding response factors $f_{j,A}$ and $f_{j,B}$, wherein j denotes the individual detectors.

$$x_{i,j} = m_i \left(w_A f_{A,j} + w_B f_{B,j} \right) \qquad (4.17)$$

The weight fractions w_A and w_B of the monomers can be calculated using

$$\frac{1}{w_A} = 1 - \frac{\left(\dfrac{x_1}{x_2} * f_{2,A} - f_{1,A}\right)}{\left(\dfrac{x_1}{x_2} * f_{2,B} - f_{1,B}\right)} \qquad (4.18)$$

Once the weight fractions of the monomers are known, the correct mass of polymer in the slice can be calculated using

$$m_i = \frac{x_i}{w_A * \left(f_{1,A} - f_{1,B}\right) + f_{1,B}} \qquad (4.19)$$

and the molar mass M_C of the copolymer is obtained by interpolation between the calibration lines of the homopolymers [35].

$$\ln M_C = \ln M_B + w_A * (\ln M_A - \ln M_B) \qquad (4.20)$$

wherein M_A and M_B are the molar masses of the homopolymers, which would elute in this slice of the peak (at the corresponding elution volume V_e).

It is clear that the interpolation between the calibration lines cannot be applied to mixtures of polymers (polymer blends): If the calibration lines are different, different molecular weights of the homopolymers will elute at the same volume. The universal calibration is not capable of eliminating these errors, either, which originate from the simultaneous elution of two polymer fractions with the same hydrodynamic volume, but different composition and molecular weight. Ogawa [36] has already shown by using a simulation technique that the molecular weights of polymers eluting at the elution volume V_e are given by the corresponding coefficients K and α in the Mark-Houwink equation.

In SEC of a polymer blend, molecular weights of the homopolymers eluting in the same interval can be calculated using

$$\ln M = \frac{A * V_e}{1+a} + \frac{B - \ln K}{1+a} \qquad (4.21)$$

The architecture of a copolymer (random, block, graft) has also to be taken into account, as Revillon [37] has shown by SEC with RI, UV, and viscosity detection. Intrinsic viscosity varies largely with molar mass according to the type of polymer, its composition, and the nature of its components. Tung [38] found that for block copolymers in good SEC solvents the simpler first approach (Eq. 4.20) is more precise.

Obviously it is feasible to use a combination of molar mass sensitive detectors, such as light scattering and viscosity detector with two concentration detectors (such as UV and RI), from which the (average) composition for each fraction can be obtained, and thus the amount of polymer in the fraction [39–42].

When multiple detection is used, one has also to be aware of errors arising from peak spreading between the detectors (and from inaccurate shift time). Bielsa and Meira [43] have studied the influence on instrumental broadening in copolymer analysis with dual-detection SEC, and demonstrated the effect of different corrections. A highly important question in multiple detection is the arrangement of detectors: whether in series or parallel. In order to minimize peak dispersion between the detectors, several authors have used a 1:1 split of the solvent stream and arranged two detectors in each branch (e.g. MALLS+RI and UV+DV). The main problem in such a setup is, however, that one must make sure that the split ratio is absolutely constant over the entire chromatogram, which is neither easily achieved nor measured with sufficient precision. The split ratio is typically controlled by a back pressure regulator. It is clear that the back pressure in the branch with the viscometer will change, when a polymer fraction enters its capillaries. Concentration errors may also influence the reliability of the results [44]. Another source of error in SEC is, of course, the separation itself, which may be influenced by various effects [45].

4.4 Analysis of Homopolymers by SEC

The easiest case in SEC is the determination of the MMD of a homopolymer, for which a sufficient number of calibration standards is available. Such polymers are PS, PMMA, PEG etc.. Provided that the chromatographic system works well and an appropriate mobile phase is used, there remains the problem of a reliable calibration, which must be established and evaluated very carefully. The importance of the quality of the calibration shall be demonstrated for PS and PMMA.With these polymers, the influence of the order of the polynomial fit on the quality of the calibration lines can be evaluated. This procedure can improve the reliability of the MMDs determined by SEC considerably.

4.4.1 Calibration of the Chromatographic System

Aim

SEC is capable of providing the number average as well as the weight average of molar mass, M_n and M_w, from one single measurement. The most common approach uses the peak position calibration (with narrow MMD standards), from which a calibration function is obtained. The accuracy of the results depends strongly on the number of standards, the molar mass range covered by them, and the order of the fit.

Calibration Standards. Narrow-disperse PS from Waters and poly(methyl methacrylate) from Polymer Laboratories

Materials

Chromatographic System. Modular system comprising of a JASCO 880 PU pump equipped with a Rheodyne injector

Equipment

Columns. Phenogel M, average particle size 5 µm, column size 600x7.6 mm

Mobile Phase. Chloroform (stabilized with 2-methyl-butene), and 2-butanone, both HPLC grade

Detectors. Density detection system DDS 70 (Chromtech, Graz, Austria) coupled with an ERC 7512 RI detector

Column Temperature. 25 °C

Sample Concentration. 1.0–4.0 g/L

Injection Volume. 50 and 100 µL

As described in Section 4.3, a reliable calibration for each column system has to be established and its validity evaluated. When using the peak position calibration, one should, however, always make sure that the correct fit order is used. As can be seen from Fig. 4.3, a first order fit (which is obtained by linear regression) does not represent the actual calibration function, which seems to be the case for a third- or fourth-order polynomial.

Preparatory Investigations

One of the most important sources of error in establishing a calibration is the number of data points. When calibrating an identical column as above for PMMA (with 6 standards of M=1680–58500), one can improve the reliability of the calibration function considerably by using the internal standard (THF) as an additional data point. While a third-order fit works quite well in both cases, a fourth-order fit would produce a grossly inaccurate fit, as shown in Fig. 4.4.

The appropriate fit order can be found easily by calculating the molar mass averages for all standards. In Table 4.6, the results obtained with different fit order are given for PS 50000. The best

Evaluation

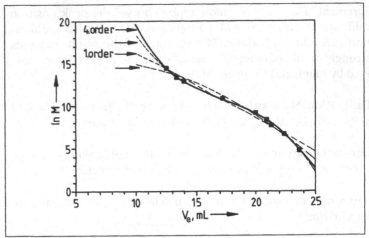

Fig. 4.3. Calibration functions for polystyrene in butanone, as obtained for a Phenogel M column (5 µm, 600x7.6 mm) with different fit order

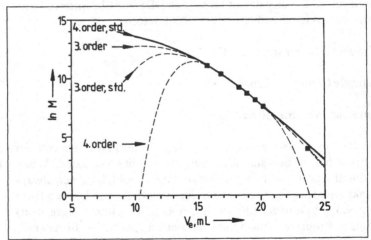

Fig. 4.4. Calibration curves for PMMA on Phenogel M (60 cm) in chloroform at 25 °C, as obtained with polynomial fit of different order

agreement with the specification given by the distributor is obtained by a fourth-order fit.

In Fig. 4.5, the number and weight distribution of PS 50000 under these conditions is shown together with the calibration function used. It is recommended to use such a representation, because it shows the quality of the calibration function. The fit parameters A-E (from Eq. 4.7) should also be given with the report of the analysis.

Table 4.6. Molar mass averages, as obtained for PS 50000 on a Phenogel M column (5 μm, 600x7.6 mm) with different fit order

	1. order	2. order	3. order	4. order
M_w	40900	66800	46000	49300
M_n	39100	62600	44400	47700
M_w/M_n	1,072	1,058	1,037	1,033

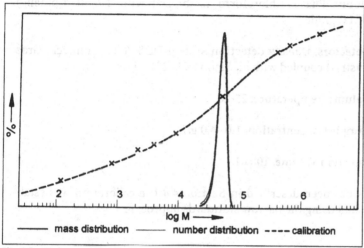

Fig. 4.5. Molar mass distribution of PS 50000, stationary phase: Phenogel M, mobile phase: butanone

4.4.2 Analysis of a Polystyrene Hompolymer

As PS is a common polymer that is used for many applications, Aim reliable methods for its molecular characterization are required. Typically, styrene is polymerized via a free radical mechanism; hence the MMD is expected to be much wider than the polymers prepared by anionic polymerization, and the polydispersity of such polymers will be approximately 2.

Crosslinked PS packings are usually applied in SEC of polymers. Depending on the detector used, the following mobile phases are common: toluene and butanone can be used with RI and density detection, but not with UV detection, THF and $CHCl_3$ are possible with all detectors. The ELSD may also be applied, but with caution: because its linear range is small, and the molar mass dependence of its response is still unclear.

Materials   **Calibration Standards.** Narrow-disperse PS from Waters.

                     Polymer. PS was prepared by suspension polymerization at 80 °C
                     with dibenzoyl peroxide as initiator.

Equipment   **Chromatographic System.** Modular system comprising of a
                     Gynkotek 300C pump equipped with a VICI injector.

                     Columns. Phenogel M, average particle size 5 μm, column size
                     600x7.6 mm

                     Mobile Phase. Chloroform (stabilized with 2-methyl-butene),
                     HPLC grade

                     Detectors. Density detection system DDS 70 (Chromtech, Graz,
                     Austria) coupled with an Sicon LCD 201 RI detector.

                     Column Temperature. 25 °C

                     Sample Concentration. 1.0–4.0 g/L

                     Injection Volume. 100 μL

Preparatory   As has been described in Section 4.4.1, a calibration was estab-
Investigations   lished using the narrow standards approach.

Measurement   Polystyrene was dissolved in the mobile phase from the reservoir
                     (in a 10 μL flask) and 10 μL ethanol were added as internal stan-
                     dard for flow rate correction. After a dissolution time of at least
                     20 min the sample was injected. From each of the detectors, (den-
                     sity and RI), the MMD can be calculated. The results obtained
                     from the RI data are shown in Fig. 4.6.

Evaluation   The MMD of PS can be determined by SEC on crosslinked PS
                     phases (from various manufacturers) in chloroform, THF, or
                     butanone as the mobile phase. Depending on the mobile phase,
                     UV, RI, and density detection can be applied. SEC calibration is
                     performed best with narrow standards. A sufficient number of
                     standards should be used (6–10), the molar masses of which
                     should cover a reasonable range. It is advantageous, to select
                     standards in such a way that they will elute in similar volume
                     increments.

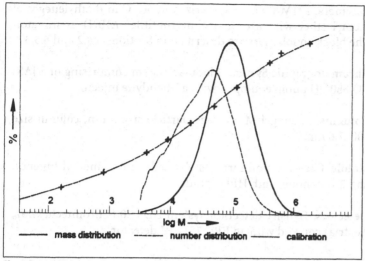

Fig. 4.6. MMD of a polystyrene prepared by suspension polymerization, stationary phase: Phenogel M, mobile phase: chloroform

4.5 Analysis of Copolymers by SEC with Dual Detection

In the analysis of copolymers, there is one more important parameter to be determined besides the weight fraction of each individual molar mass: the chemical composition of the polymer, which may vary within the peak. In this case, no reliable results can be expected with just one concentration detector if the detector monitors both components (monomer units) with different sensitivities. If, however, two detectors are coupled, one may utilize the different sensitivities to obtain information about chemical composition at any point of the MMD.

Characterization of copolymers requires a determination of the MMD and CCD. Using SEC, one may achieve a separation according to hydrodynamic volumes, which can be correlated to molar mass. The chemical composition along the polymer peak can be determined by dual detector SEC. This technique yields not only the overall chemical composition. It can provide additional information on the presence of homopolymers in block or graft copolymers, even though it is not fully equivalent to a two-dimensional separation.

Aim

Calibration Standards. Narrow-disperse PS from Waters, poly(methyl methacrylate) from Polymer Laboratories, poly(decyl methacrylate) from Röhm.

Materials

Polymers. PMMA was polymerized at 60 °C in diethyleneglycol diethyl ether with azo-bis-isobutyrontitile (AIBN) as initiator. The block copolymers are described in Sections 6.5.2 and 6.5.3

Equipment

Chromatographic System. Modular system comprising of a JASCO 880 PU pump equipped with a Rheodyne injector.

Columns. Phenogel M, average particle size 5 μm, column size 600x7.6 mm

Mobile Phase. Chloroform (stabilized with 2-methyl-butene), and 2-butanone, both HPLC grade

Detectors. Density detection system DDS 70 (Chromtech, Graz, Austria) coupled with an ERC 7512 RI detector.

Column Temperature. 25 °C

Sample Concentration. 1.0–4.0 g/L

Injection Volume. 50 and 100 μL

Preparatory
Investigations

The determination of chemical composition by dual detector SEC requires the determination of the response factors of both detectors for the homopolymers (Section 4.3). Moreover, the SEC calibrations for both homopolymers have to be established if the corresponding standards are available. The universal calibration may be applied, provided that the Mark-Houwink parameters are known with sufficient accuracy. This is, however, seldom the case for those polymers, for which no standards with a narrow MMD are available. The determination of response factors need not be discussed in detail for high molar masses. The special case of low molecular samples will be the subject of the next section.

Much more critical are the SEC calibrations: as can be seen from Fig. 4.7 the calibration lines for PS and PMMA are considerably different, especially in the low-molecular region. In both cases, a fourth-order polynomial fit works quite well.

If the PS calibration is applied to PMMA, considerable errors in molar mass will be the consequence.

With dual detection, one may calculate the chemical composition of a polymer. Once the composition is known, one may interpolate between the calibration lines. The effect of such a procedure is given in Table 4.7 and Fig. 4.8: A PMMA sample (prepared by free-radical polymerization) was analysed by SEC with coupled density and RI detection. Three different approaches were used in the calculation of molar mass averages:

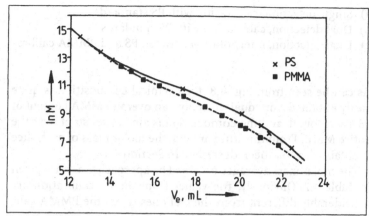

Fig. 4.7. SEC calibration functions for PS and PMMA (both with a fourth-order polynomial fit), stationary phase: Phenogel M (600x7.6 mm), mobile phase: 2-butanone

Table 4.7. Comparison of molar mass averages of a PMMA sample, as obtained on Phenogel M in 2-butanone with different calibrations

Method	M_w	M_n
Single detection, PS calibration	59400	37100
Dual detection, PS calibration	59400	37100
Dual detection, PS and PMMA calibration	38600	19100

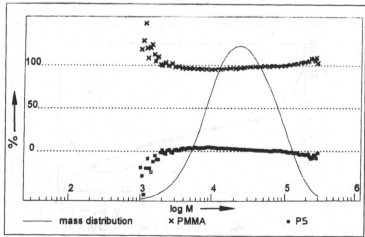

Fig. 4.8. Molar mass distribution and chemical composition of a PMMA sample, stationary phase: Phenogel M, mobile phase: 2-butanone, detector: coupled density and RI

a) Single detection, calibration with PS standards
b) Dual detection, calibration with PS standards
c) Dual detection, interpolation between PS and PMMA calibration.

As can be seen from Fig. 4.8, the chemical composition is quite easily obtained from dual detection: an overall PMMA content of 98.4% is found, and the composition is almost constant over the entire MMD. From this information, the molar mass of each slice is obtained, as has been described in Section 4.3.

The molar mass averages obtained by approaches a–c are given in Table 4.7. Obviously, the values from the PS calibration are considerably different from the real ones (from the PMMA calibration).

Before analysing copolymers, one should always start with the analysis of the homopolymers: If the response factors for both are reliable, one should find the compositions 0 and 100%, repectively, in both cases (as shown in Fig. 4.8).

For PMMA and poly(decyl methacrylate) (PDMA) the difference between the calibration lines is much smaller: all the standards seem to fall on one line. In the case of PDMA, however, the order of the fit is the crucial point. Because of the limited number of avaliable standards, a too high order produces erroneous results, as can be seen fom Fig. 4.9.

Measurement   An analysis of copolymers by SEC with dual detection will require the following steps:

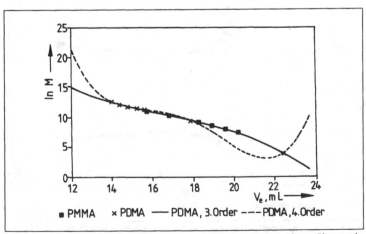

Fig. 4.9. Calibration functions for PMMA and PDMA, stationary phase: Phenogel M, mobile phase: 2-butanone

1. Determination of SEC calibration functions for both homopolymers, preferably with narrow MMD standards, if these are available.
2. Determination of the response factors of both detectors for the homopolymers. In the case of low molar mass samples, the dependence of response factors on molar mass has also to be considered.
3. Determination of interdetector volume.
4. Analysis of the corresponding homopolymers. From the results the accuracy of steps 1–3 can be evaluated.
5. Analysis of the copolymer samples.

A block copolymer of styrene and methyl methacrylate, which had been synthesized by anionic polymerization with subsequent addition of monomers is separated by SEC. With the same conditions as above, the chemical composition of each slice of the peak could be obtained from coupled density and RI detection, as shown in Fig. 4.10.

The analysis of homopolymer standards prior to copolymer analysis is very important, because it provides information on: Evaluation

– the quality of the calibration functions for both homopolymers
– the accuracy of response factors (the suitability of the mobile phase)

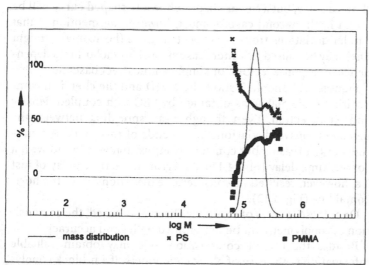

Fig. 4.10. MMD and chemical composition of a block copolymer PMMA-b-PS determined by dual detection SEC, stationary phase: Phenogel M, mobile phase: 2-butanone

– the accuracy of the interdetector volume
– the extent of peak dispersion between the detectors.

Once the quality of the calibration functions has been evaluated as described above, the response factors for the homopolymers should be sufficiently different. If this is not the case, a change of the mobile phase may bring about a considerable improvement. One should, however, avoid mixed mobile phases, because in this case preferential solvation of the polymer may affect the accuracy of results. It is evident that erroneous response factors will yield overall compositions different from 0 and 100%, respectively, for the homopolymers.

As chemical composition is obtained in dual detection SEC from the ratio of the detector signals (x_1/x_2) along the chromatogram, two other sources of error affect narrow peaks (such as monodisperse oligomers) much stronger than broad MMD polymers. Obviously, an *incorrect interdetector volume* and *peak dispersion* will affect the results, but in a different way. An incorrect interdetector volume will shift the second detector tracing against the first one. An insufficiently compensated delay will lead to a high value of x_1/x_2 at the beginning of the peak and a low value at the end. This will result in an apparent variation of chemical composition along the peak. Peak dispersion between the detectors, however, leads to a low ratio (x_1/x_2) on both sides of the peak. Especially at the very ends of narrow peaks, where both x_1 and x_2 approach zero, this will cause considerable errors in the chemical composition. In the first case, the composition shows a slope, but a linear shape, while a U-shaped curve will be found in the second case. It must, however, be mentioned that similar deviations from the expected shape (horizontal, straight line) may be caused by other reasons, such as molar mass dependence of response factors, or simply an incorrect baseline.

Figures 4.11 and 4.12 show the MMD and the chemical composition of PS 50000, as obtained by SEC with coupled density and RI detection. Even though peak spreading between the detectors caused deviations at the ends of this extremely narrow peak, a reasonably constant composition was found with a correct time delay (Fig. 4.11). An error in the time delay of just 1 s, however, resulted in considerable deviations from the horizontal line (Fig. 4.12).

Once these sources of error have been eliminated, the composition of copolymers can be determined with good accuracy.

Besides the overall composition, one may obtain valuable information in the case of block copolymers. If a n-block copolymer contains homopolymer(s) or the (n-1)-block copolymer, a shoulder (or a second maximum) in the MMD will be found that

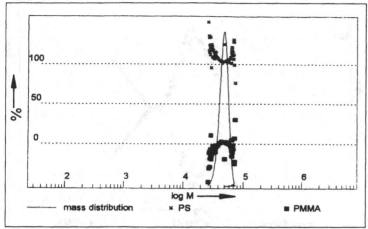

Fig. 4.11. MMD of PS 50000 (calibrations: PS and PMMA)

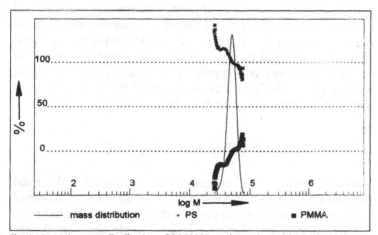

Fig. 4.12. Molar mass distribution of PS 50000 as above, but with an incorrect time delay between the detectors (1 s too low)

should have a different composition. Figure 4.13 shows the MMD of a block copolymer of methyl methacrylate and decyl methacrylate, in which a second (lower molecular) maximum can be identified by dual detection as poly(decyl methacrylate). The increase of the PDMA content with molar mass in the main fraction is quite reasonable and corresponds well with what should be expected from the synthesis. These findings can be supported by liquid chromatography under critical conditions (LCCC) for both of the blocks [46].

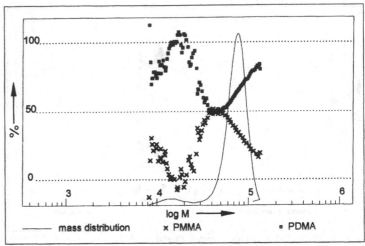

Fig. 4.13. MMD and chemical composition of a PMMA-PDMA block copolymer, as obtained by SEC with density and RI detection

Obviously, a combination of LCCC and dual detector SEC will give a much better understanding of a copolymer than each of these techniques, when applied alone. SEC yields the composition as a function of molar mass, but does not give information on whether a peak is a copolymer or a mixture of co-eluting homopolymers. LCCC yields the length of the blocks, but does not provide information, whether they are linked together.

4.6 Analysis of Oligomers by SEC

Response factors of SEC detectors may depend on molar mass not only for the lowest oligomers, but even up to molar masses of several thousands. In some cases, such a compensation has to be performed even for higher polymers. Neglecting this dependence can lead to severe errors [47]. In these investigations, ethoxylated fatty alcohols (FAE) were analysed using SEC in chloroform. While chromatograms looked normal in density detection, the response for the lower oligomers changed its sign in RI detection.

4.6.1 Analysis of Fatty Alcohol Ethoxylates

Aim

Fatty alcohol ethoxylates (FAE) are of widespread use as non-ionic surfactants. Since FAE do not contain UV-absorbing groups, they can only be detected by bulk property detectors. (The ELSD suffers from the fact that its signal is influenced by the surface tension of the eluate, which will vary with the number of EO units). The analysis of these products is complicated by the fact that their synthesis is based on fatty alcohols, which are typically not 100% pure. This means that they may consist of several polymer homologous series (with varying number of oxyethylene units) with different alkyl end groups.

If a pure alkanol has been used in the synthesis, or if the individual homologous series have been separated by preparative LCCC (see Chapter 6), one may use a correction using Eq. 4.19 to obtain correct MMD. If the amount of the other homologous series is not too high, this correction will work even with the unfractionated products.

Calibration Standards. Narrow-disperse polyethylene glycols from Polymer Laboratories, fatty alcohols and the corresponding ethoxylates (Brij) from FLUKA — Materials

Oligomers. Diethylene glycol monoalkyl ethers were purchased from FLUKA, higher monodisperse oligomers were supplied by H.-R. Holzbauer (Institute for Applied Chemistry Adlershof, Berlin, Germany).

Chromatographic System. Modular system comprising of a JASCO 880 PU pump equipped with a Rheodyne injector. — Equipment

Columns. Phenogel M, 5 µm average particle size, column size 600x7.6 mm

Mobile Phase. Chloroform (stabilized with 2-methyl-butene), and 2-butanone, both HPLC grade

Detectors. Density detection system DDS 70 (Chromtech, Graz, Austria) coupled with an ERC 7512 RI detector.

Column Temperature. 25 °C

Sample Concentration. 3.0–10.0 g/L

Injection Volume. 100 µL

As the first step, the response factors of PEG, the alkanols and the monodisperse oligomers were determined for both detectors. In Fig. 4.14, a plot of the response factors vs 1/M is shown. The monodisperse oligomers fall on the lines connecting the points of high molecular PEG 6000 and the corresponding alkanols. — Preparatory Investigations

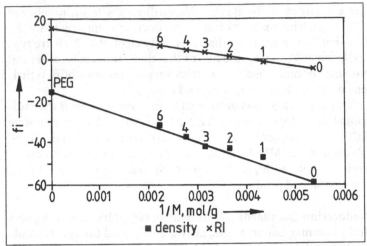

Fig. 4.14. Response factors of FAE in density and RI detection: monododecyl ethers of PEG (0–6 EO units) in CHCl₃

Once the response factors of sufficiently high-molecular PEG (with M >4–6000) and the corresponding alkanols have been determined, the response factors can be calculated for all oligomers of each polymer homologous series (from the intercept f_∞ and the slope K in Eq. 4.16). It is also important to use the appropriate SEC calibration for each individual homologous series instead of that obtained with PEG standards. In most cases, it will be sufficient to use the alkanol, the (sufficiently resolved) lower oligomers of an appropriate sample, and PEG as the upper end of the calibration.

Measurement   The analysis of FAE requires the knowledge of the response factors and their dependence on molar mass. This becomes very clear from Fig. 4.15, which shows a chromatogram of Brij 30, which was obtained with density and RI detection.

Evaluation   It is evident that the correction using slope and intercept will only work with the density detector output, because the RI output changes its sign within the peak, as could be expected from Fig. 4.14. The correct MMD may be calculated from the density tracing, as is shown in Fig. 4.16. Without compensation of the molar mass dependence of response factors, the low-molar mass end of the MMD would be overestimated.

There is, however, another approach that works as well: If one considers FAE as block copolymers consisting of the fatty alcohol

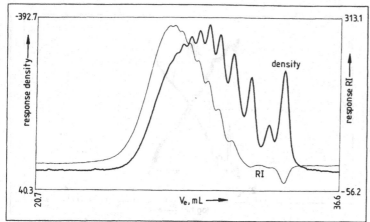

Fig. 4.15. SEC of Brij 30 (ethoxylated dodecanol), stationary phase: 4 columns Phenogel (2x100 Å, 2x500 Å), mobile phase: chloroform

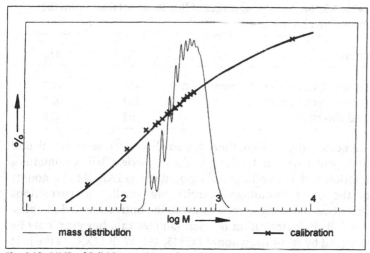

Fig. 4.16. MMD of Brij 30, experimental conditions see Fig. 4.15

and a PEG block (without end groups), one may use dual detection to determine the composition of the sample, and thus the correct response factors. Both approaches are equivalent, as has been shown in the same paper [47].

Figure 4.17 shows the MMD and chemical composition of Brij 30, as obtained from dual detection. The corresponding molar mass averages are given in Table 4.8.

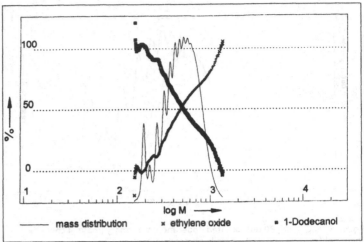

Fig. 4.17. MMD and chemical composition of Brij 30, as obtained by SEC with coupled density and RI detection

Table 4.8. Molar mass averages of Brij 30, as obtained from the chromatogram in Fig. 4.15

Method	M_w	M_n
Density detection, no compensation	458	380
Density detection, slope	504	423
Dual detection	502	418

As can easily be seen, there is a small peak close to the alkanol peak, which does not belong to the C_{12} series: Brij 30 contains a fraction with C_{14} and a small amount of C_{16} FAE. If the amounts of the other homologous series are small, the corrections described above work well even with the unfractionated samples.

A full characterization of such samples can, however, only be achieved by two-dimensional LC [48, 49]. With LCCC as the first and SEC as the second dimension a three-dimensional map of such a product can be obtained [50, 51]. In some cases, even the one-dimensional approach may provide most of the required information, as can be seen from the following example.

4.6.2 Analysis of Block Copolymers of Ethylene Oxide and Propylene Oxide

Aim

Block copolymers of ethylene oxide (EO) and propylene oxide (PO) are useful for many applications: The low temperature dependence of their viscosity makes them attractive as lubricants (because of their low toxicity, even in food industry). As amphiphilic molecules they may form micelles, which can be utilized in pharmacology and many other fields.

Reliable characterization of these materials is important, but also complicated. The problems in the characterization of EO-PO two-block copolymers are basically the same as with FAE. Even more interesting materials are the three-block copolymers, which may contain not only the homopolymers PEG and PPG, but also two-block copolymers. These materials are typically prepared either by ethoxylation of PPG or by propoxylation of PEG, mostly with basic catalysts. Under these conditions, chain-transfer reactions may occur that lead to the formation of a homopolymer besides the copolymer. Since this side product has a considerably lower molar mass than the desired copolymer, it can be separated from the main fraction by SEC and identified by dual detection.

Calibration Standards. Narrow-disperse PEG from Polymer Laboratories, polydisperse PEG and PPG from FLUKA — Materials

Oligomers. Oligoethylene glycols (n=2–6) and oligopropylene glycols (n=2–3) from FLUKA, block copolymers were supplied by H.-R. Holzbauer (Institute for Applied Chemistry Adlershof, Berlin, Germany).

Chromatographic System. Modular system comprising of a JASCO 880 PU pump equipped with a Rheodyne injector. — Equipment

Columns. Phenogel M, 5 µm average particle size, column size 600x7.6 mm

Mobile Phase. chloroform (stabilized with 2-methyl-butene), and 2-butanone, both HPLC grade

Detectors. Density detection system DDS 70 (Chromtech, Graz, Austria) coupled with an ERC 7512 RI detector.

Column Temperature. 25 °C

Sample Concentration. 3.0–10.0 g/L

Injection Volume. 100 µL

As the first step, the response factors of PEG, PPG, and the available monodisperse oligomers were determined for both detectors, as has been described above. With the same standards, SEC — Preparatory Investigations

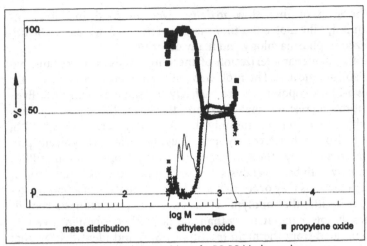

Fig. 4.18. MMD and chemical composition of a EO-PO block copolymer , as obtained from SEC with coupled density and RI detection

calibrations for PEG and PPG were established that were found to be considerably different.

Measurement

Once the necessary calibrations had been performed, block copolymers were analysed as decribed above. In Fig. 4.18, the MMD of a block copolymer is shown that was obtained by propoxylation of PEG 600. This sample contains a fraction with lower molar mass that was assumed to be PPG formed by chain transfer to water in the reaction mixture. From the signals of the density and the RI detector, the chemical composition was determined at any point of the MMD, which gave the proof that the low molar mass fraction was indeed pure PPG.

References

1. CRAVEN JR, TYRER H, POK LAI LI S, BOOTH C, JACKSON D (1987) J Chromatogr 387: 223
2. MORI S, NISHIMURA S (1993) J Liq Chromatogr 16: 3359
3. TUNG LH (1969) J Appl Polym Sci 13: 775
4. YAU WW, KIRKLAND JJ, BLY DD (1979) Modern Size Exclusion Chromatography. Wiley, New York, p 332
5. LEDERER K, BEYTOLLAHI-AMTMANN Y, BILLIANI J (1994) J Appl Polym Sci 54: 47
6. KILZ P (1995) Prep Pod EAS Symp, page 135
7. YAU WW KIRKLAND JJ, BLY DD (1979) Modern Size Exclusion Chromatography. Wiley, New York, p 64
8. BEREK D, PLEHA T, PEVNA Z (1976) J Chromatogr Sci 14: 560
9. SPYCHAJ T, LATH D, BEREK D (1979) Polymer 20: 437

10. TRATHNIGG B, YAN X (1993) J Chromatogr 653: 199
11. JOHANN C, KILZ P (1989) Prep 1st Int Conf Mol Mass Charact Polym, Bradford, UK
12. TRATHNIGG B, THAMER D, YAN X (1995) Int J Polymer Anal Character 1: 35
13. BALKE ST (1984) Quantitative Column Liquid Chromatography, J Chromatogr Libr 29: 204
14. YAU WW, KIRKLAND JJ, BLY DD (1979) Modern Size Exclusion Chromatography. Wiley, New York, p 294
15. BALKE ST (1984) Quantitative Column Liquid Chromatography, J Chromatogr Libr 29: 214
16. BALKE ST, HAMIELEC AE, LeCLAIR BP, PEARCE SL (1969) Ind Eng Chem Prod Res Dev 8: 54
17. McCRACKIN FL (1977) J Appl Polym Sci 26: 191
18. GRUBISIC Z, REMPP P, BENOIT, H (1967) J Polym Sci B 5: 573
19. TSITSILLIANIS C, DONDOS A (1990) J Liq Chromatogr 13: 3027
20. KURATA, M, TSUNASHIMA, Y, IWAMA, M, KAMADA, K (1975) In: Brandrup J, Immergut I (eds), Polymer Handbook. Wiley, New York, p IV-1
21. GLÖCKNER G (1991) Gradient HPLC of Copolymers and Chromatographic Cross-Fractionation. Springer, Berlin Heidelberg New York, pp 28 ff
22. BALKE ST (1984) Quantitative Column Liquid Chromatography, J Chromatogr Libr 29: 223
23. CANDAU F, FRANCOIS J, BENOIT H (1974) Polymer 15: 626
24. SCHULZ GV, HOFFMANN M (1957) Makromol Chem 23: 220
25. KOBATAKA Y, INAGAKI, H (1960) Makromol Chem 40: 118
26. GÉCZY, I (1972) Tenside Deterg 9: 117
27. CHENG, R, YAN, X (1989) Acta Polym Sin 12: 647
28. NAKAHARA H, KINUGASA S, HATTORI S, MUKOUYAMA M, HAYASHI T (1993) Bunseki Kagaku 42: 273
29. TRATHNIGG, B, YAN, X (1993) J Appl Polym Sci Appl Polym Symp 52: 193
30. MORI S, SUZUKI T (1981) J Liq Chromatogr 4: 1685
31. TRATHNIGG, B (1990) J Liq Chromatogr 13: 1731
32. TRATHNIGG, B (1991) J Chromatogr 552: 505
33. TRATHNIGG, B, YAN X (1992) Chromatographia 33: 467
34. HOPIA AI, OLLILAINEN VM (1993) J Liq Chromatogr 16: 2469
35. RUNYON JR, BARNES DE, RUDEL JF, TUNG LH (1969) J Appl Polym Sci 13: 2359
36. OGAWA T (1979) J Appl Polym Sci 23: 3515
37. REVILLON A (1980) J Liq Chromatogr 3: 1137
38. TUNG LH (1979) J Appl Polym Sci 24: 953
39. GARCIA-RUBIO LH (1987) In: Provder T (ed) Detection and Data Analysis in Size Exclusion Chromatography. ACS Symposium Series 352, American Chemical Society, New York, p220

40. GORES, F, KILZ, P (1993) In: Provder T (ed) Chromatography of Polymers. ACS Symposium Series 521, American Chemical Society, New York, p 122
41. MOUREY TH, BALKE ST (1993) In: Provder T (ed) Chromatography of Polymers. ACS Symposium Series 521, American Chemical Society, New York, p. 180
42. GORES F, KILZ P (1993) In: Provder T (ed) Chromatography of Polymers. ACS Symp. Ser. 521, ACS, New York, p 123
43. BIELSA RO, MEIRA GR (1992) J Appl Polym Sci 46: 835
44. GUAITA M, CHIANTORE O (1993) J Liq Chromatogr 16: 633
45. BARTH HG (1987) In: Provder T (ed) Detection and Data Analysis in Size Exclusion Chromatography. ACS Symposium Series 352, American Chemical Society, Washington, DC, p 29
46. PASCH H, TRATHNIGG B, AUGENSTEIN M (1994) Makromol Chem Phys 195: 743
47. TRATHNIGG, B THAMER D, YAN X, MAIER B, HOLZBAUER, H-R, MUCH, H (1993) J Chromatogr 657:365
48. SCHULZ, G, MUCH, H, KRÜGER, H, WEHRSTEDT, C (1990) J Liq Chromatogr 13:1745
49. PASCH, H, KRÜGER, H, MUCH, H, JUST, U (1992) J Chromatogr 589:289
50. TRATHNIGG, B, KOLLROSER, M (1995) Int J Polymer Analysis Characterization 1, 301
51. TRATHNIGG B (1995) GIT Fachz Lab 39:

5 Liquid Adsorption Chromatography

5.1 Peculiarities

Adsorption chromatography is based on the retention of solute molecules by surface adsorption. Intermittent capture and release of solute molecules by the stationary phase are controlled by two basically different mechanisms or some combinations thereof. In regard to adsorption-desorption phenomena, an abrupt process is the critical step leading to sorption or desorption. This process is typified by molecular desorption from surfaces where molecules can detach, and then do so suddenly, if they possess sufficient activation energy to cause the necessary rearrangement or rupture of chemical or physical bonding. Quite different in effect are the diffusion-controlled sorption-desorption kinetics where a change occurs only gradually as molecules diffuse in and out of localized regions [1].

The behaviour of synthetic polymer molecules deviates in several respects from the behaviour of low-molar-mass compounds. The differences are caused by the following properties of polymers [2]:

- small diffusion coefficients of macromolecules in solution,
- size of the macromolecules, which may be of the same magnitude as the pores of the packing,
- retention of polymers via "trains" of numerous repeat units,
- the flexibility of chain molecules, which enables conformational changes to occur,
- limited solubility of polymers.

The interactions between flexible macromolecules in solution and the surface of the packing depend on the magnitude of the adsorption energy. The beginning of an adsorption process is schematically represented in Fig. 5.1.

In a first step, a part of the repeat units of the polymer chain is bound to the active groups of the stationary phase. If the adsorption energy per repeat unit is too low, the macromolecule is repulsed from the surface. When the adsorption energy exceeds a certain limit, the macromolecule is adsorbed. If the change in adsorption enthalpy ΔH is higher than the entropy losses ΔS,

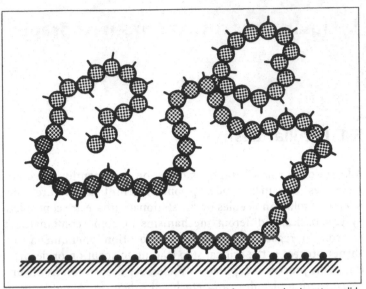

Fig. 5.1. Schematic representation of the adsorption of a macromolecule onto a solid surface (Reprinted from Ref. [3] with permission)

related to the transformation from a three-dimensional coil to a two-dimensional structure, the macromolecule changes its conformation and becomes fully adsorbed with all repeat units. Based on extensive experimental investigations using viscometry, ellipsometry, IR spectroscopy and calorimetry, the structure of the adsorbed layer of a polymer may be described using the models given in Table 5.1 [3].

The type of the final conformation of the macromolecule at the surface of the stationary phase depends on the flexibility of the polymer chain, its chemical structure, the molar mass and the number of adsorbed molecules. For example, if interacting end groups are present in the macromolecule, a brush-type adsorption layer may be obtained. Most types of macromolecules are flexible and random coil conformation is typical. In dilute solutions, these coils contain much solvent and the polymer segments are mobile. Special conformations may be favoured by external forces. Under the conditions of HPLC, random coils may be deformed in a way that extended trains of segments come in close contact with the surface of the stationary phase and only a few loops and free chain ends protract into the mobile phase. In pores, coiled macromolecules may adjust their conformation to the size of the cavity in order to gain the maximum of interaction enthalpy.

In interaction chromatography the strength of the eluent has a dramatic effect on retention. With a weak eluent, the reten-

Table 5.1. Model presentation of the adsorption of macromolecules

Model	Relations between molar mass and	
	surface thickness	adsorbed amount
flat layer	independent of M	independent of M
coil	independent of M	$\sim M^{0.5}$
collapsed coil	$\sim M^{1/3}$	$\sim M^{1/3}$
brush	$\sim M$	$\sim M$
loops and trains	$\sim M^a$ $0 < a < 0.5$	independent of M

tion time of a polymer usually exceeds by far any reasonable period of experimental work. Increase of the elution strength will eventually lead to a sudden change to the opposite behaviour: the polymer is not retained at all and leaves the column at dead volume V_0. A small alteration of elution conditions causes transition from zero retention to infinity. This "on or off" behaviour can be understood as a consequence of multiple attachment [4].

Synthetic polymers consist of a large number of repeat units. In principle, all of them have the chance of becoming adsorbed but the conformation of adsorbed polymer coils usually comprises loops and tails extending in the solution and trains of adsorbed repeat units. A polymer chain is retained in the stationary phase as long as one of its repeat units is adsorbed. A chain can migrate only if all constituting units are in the mobile phase. Assuming independent adsorption-desorption equilibrium for each unit, the mobility condition of a macromolecule is a function of the corresponding probabilities of the repeat units and the chain length. For weak eluents and long polymer chains it can be assumed that there is always a repeat unit interacting with the packing material. Accordingly, the probability is very high that the macromolecule is retained for a very long time.

Another important aspect of adsorption chromatography of polymers is solubility. Solubility in general requires a negative change in Gibbs free energy on mixing. This is easily fulfilled with low-molar-mass solutes due to the large entropy contribution. In low-molecular mixing processes, the contribution $T\Delta S_m$ is so large that even positive values of ΔH_m do not prevent dissolution. In contrast to low-molar-mass compounds, macromolecules normally have low order in the solid state. In addition to this, the regular arrangement of the repeat units along the polymer chain remains on dissolution. Furthermore, in solutions of equal weight concentrations the number of solute particles is much greater in low-molar-mass than in polymer systems. Thus, an entropy contribution to ΔG_m will be small and the condition $\Delta G_m < 0$ requires a negative change in enthalpy, $\Delta H_m < 0$ [2]. Accordingly, the solubility parameters of the solvent must be very close to that of the polymer. In addition to sufficient solubility of the macromolecules in the eluent, a certain solvent strength of the eluent is required. In view of the separation mechanism in adsorption chromatography, the solvent strength must be high enough to promote desorption of the macromolecules from the packing. The solvent strength, however, must not be too high, otherwise a proper adsorption-desorption equilibrium cannot be established. If thermodynamically good solvents are used as the eluent, adsorption does not occur, and on porous packings the size exclusion mode becomes operational, see Chapters 2 and 4.

In order to establish an appropriate solvent strength, mixtures of different solvents are used as eluents. The solvent strength of the mixture is a function of the properties of the components, and the eluent can be adjusted to a certain solvent strength by changing the ratio of the components. The classification of solvents with respect to solvent strength is dealt with in Section 5.2.

Polymer samples, which are homogeneous in chemical nature and chain length, can be eluted in an isocratic elution regime. In this case the composition of the eluent is constant throughout the chromatographic run. Accordingly, oligostyrenes can be separated into single oligomers on a polar silica column using n-pentane-THF 87:13% by volume as the eluent (see Fig. 5.2) [5].

Samples which consist of species differing substantially in molar mass or chemical composition cannot be separated isocratically, because the different species would exhibit very different elution behaviour. The higher molar mass oligomers would be strongly retained due to multiple attachment. The desorption of strongly retained macromolecules requires displacement by a stronger eluent. This can be achieved by gradient elution, where the eluent strength increases within the chromatographic run [6–8]. By increasing the solvent strength of the eluent stepwise, the elution behaviour of different sample components can be adjusted to the requirements of the chromatographic experiment. Principally, gradients may be produced by changing the ionic strength, the pH or the composition of the eluent, or by changing the flow rate or the temperature. However, the most common case is the change of the eluent composition.

In addition to adsorption and solubility effects, precipitation phenomena can be used in liquid chromatography of polymers. With stationary phases of high activity, proper adsorption chro-

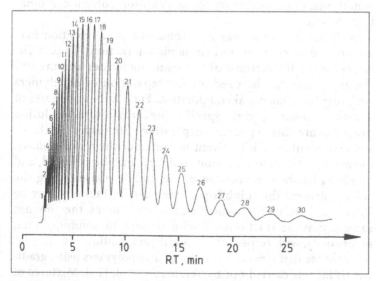

Fig. 5.2. Isocratic separation of oligomers from a polystyrene standard with an average molar mass of 2100 g/mol; stationary phase: silica gel; mobile phase: n-pentane-THF 87:13% by volume (Reprinted from Ref. [5] with permission of Hüthig & Wepf Publishers, Switzerland)

matography is possible only with polymer samples that completely elute in the course of the gradient. If the elution strength does not suffice for complete elution, a less polar packing must be chosen. With moderate interaction forces, complete elution takes place, but selective retention may become a problem. Since the starting eluent must be a rather poor solvent in order to facilitate retention, precipitation of sample components can occur upon sample injection. This effect can be used to separate copolymers with respect to chemical composition. A number of interesting applications of this high-performance precipitation liquid chromatography (HPPLC) is discussed in refs. [9–12].

The analysis of copolymers by gradient HPLC and chromatographic cross-fractionation has been extensively discussed in the excellent monograph of Glöckner [2]. In brief, the complex molar mass-chemical composition distribution of copolymers requires separation in more than one direction. The classical approach is based upon the dependence of copolymer solubility on composition and chain length. A solvent/non-solvent combination, fractionating solely by molar mass, would be appropriate for the evaluation of the MMD. To separate by chemical composition, a different solvent/non-solvent combination would be required. In general, fractionation is influenced by the molar mass and chemical composition, and the direction of separation is determined by the chromatographic conditions, in particular by the solvent/non-solvent combination chosen.

As discussed in Chapter 2, chromatographic retention may be directed by entropic and enthalpic interactions. In particular, enthalpic interactions of the solute molecules and the stationary phase may be used for the separation of copolymers with respect to chemical composition. For stationary phases of a certain polarity, very specific precipitation/redissolution processes are able to promote separation with respect to chemical composition. With solvent mixtures as the mobile phase, the precipitation/redissolution equilibria may be adjusted, and by changing the composition of the mobile phase during the elution process the solubility of the sample fractions may be changed. Thus, using gradient elution techniques, the polymer sample may be fractionated with respect to solubility and, accordingly, with respect to chemical composition.

One of the first separations of random copolymers using gradient HPLC was carried out by Teramachi et al. [13]. Mixtures of poly(styrene-co-methyl acrylate) were separated by composition on silica columns through a carbon tetrachloride/methyl acetate gradient (see Fig. 5.3).

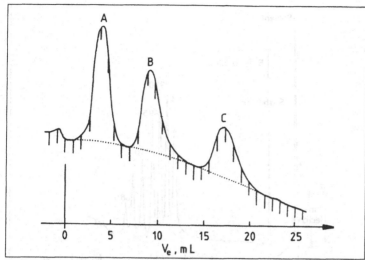

Fig. 5.3. Separation of a mixture of three random poly(styrene-co-methyl acrylate) samples by gradient HPLC: stationary phase: silica gel; mobile phase: carbon tetrachloride-methyl acetate; samples: 42% (A), 52% (B), 74% (C) methyl acrylate (Reprinted with permission from Ref. [13], Copyright 1979 American Chemical Society)

When increasing the content of methyl acetate in the eluent, retention increased with increasing methyl acrylate content in the copolymer. This behaviour fitted the normal-phase chromatographic system used. Similar separations could be achieved on other columns as well, such as polar bonded-phase columns, i.e. diol, nitrile, and amino columns [2].

Copolymers of styrene and methyl methacrylate were separated by composition in numerous eluents. Most of them represented normal-phase systems with gradients increasing in polarity and a polar stationary phase. Figure 5.4 shows the separation of a mixture of seven random poly(styrene-co-methyl methacrylate) samples on a silica column through a gradient i-octane/ (THF+10% methanol) [14].

Graft copolymerization usually yields mixtures containing the desired graft copolymer, non-grafted precursors, and homopolymers formed as a by-product during the grafting reaction. Figure 5.5 represents the separation of a graft copolymer of methyl methacrylate onto EPDM by gradient HPLC in i-octane/THF on a nitrile bonded-phase column [15]. The first peak corresponds to the non-grafted EPDM, whereas peak 3 shows the desired graft copolymer. The more polar PMMA homopolymer was retained longer and eluted in peak 4.

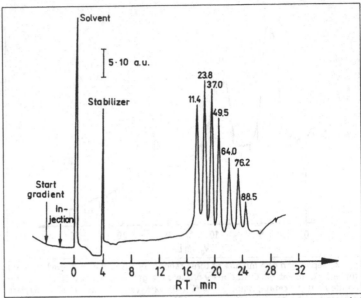

Fig. 5.4. Separation of a mixture of seven random poly(styrene-co-methyl methacrylate) samples by gradient HPLC; stationary phase: silica gel; mobile phase: i-octane-(THF+10% MeOH); samples: mass% of MMA indicated (Reprinted from Ref. [14], copyright year 1986 with kind permission from Elsevier Science-NL)

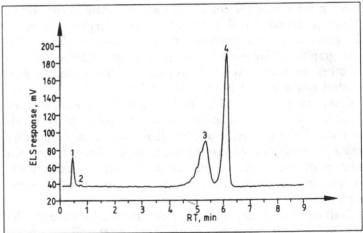

Fig. 5.5. Separation of EPDM-graft-PMMA by gradient HPLC; stationary phase: CN-bonded silica gel; mobile phase: i-octane-THF

5.2 Equipment and Materials

The typical instrument that is used for adsorption chromatography is very similar to a conventional SEC instrument (see Chapters 3 and 4). The sample is dissolved in a solvent, injected into a flow stream of the eluent at the top of the chromatographic column, and carried through the column at a constant flow rate. The columns are filled with a non-porous or porous packing material, which may interact with the solute molecules, due to a certain surface activity. Upon leaving the column, the solute molecules enter one or more sequentially attached detectors. In SEC a typical detector is a differential refractometer, but in general many different detectors may be used.

Adsorption chromatography utilizes columns that are intended to encourage adsorption and partition mechanisms. As will be shown later, at one column, depending on the composition of the mobile phase, simultaneous adsorption, partition, and size exclusion may occur. In adsorption chromatography very frequently column packings with a hydrophobic surface are used. They are termed "reversed-phase" columns, because the packings are less polar than the mobile phase.

Because of the different separation mechanisms employed in adsorption chromatography, the solvents used are usually different from the solvents in SEC. Binary or ternary mixtures of organic and aqueous solvents are frequently used as the mobile phase. The ratio of the solvents may be constant during the chromatographic run (isocratic elution) or may be programmed to vary with time (gradient elution).

Stationary Phases. In adsorption chromatography the packing material (or stationary phase) has active sites at the surface, where interaction with the solute molecules according to their polarity takes place. In general, the packing must withstand high pressure and, in gradient elution, the particle and pore size must not change when the eluent composition varies. Therefore, stationary phases of bare or modified silica are usually preferred, although cross-linked poly(acrylonitrile) or PS may be used as well [2].

Criteria for the quality of stationary phases are: size and shape of the particles, specific surface, pore size and pore size distribution, chemical stability, and pressure stability. In general, stationary phases may be classified with respect to their surface polarity (Table 5.2).

Silica gel is the most important stationary phase in interaction chromatography. Its surface behaves like a weak acid, and retention is directed by adsorption effects. The chemical modification of silica gel alters the surface by introduction of functional groups. Usually, a transformation of the polar surface to a less polar surface takes place, and retention now is directed by hydrophobic interactions. However, polar functional groups may be attached to the silica surface via a hydrophobic spacer (cyano,

Table 5.2. Classification of stationary phases in adsorption chromatography

non-polar	medium-polar	polar
styrene-divinylbenzene	cross-linked hydroxyethyl methacrylate agarose	silica gel aluminum oxide
bonded phases on silica gel n-octadecyl n-octyl phenyl	cyanopropyl diol	nitrophenyl aminopropyl

diol, amino phases). These stationary phases exhibit hydrophilic and hydrophobic properties and may be used in normal-phase as well as in reversed-phase chromatography.

Gradient elution in the normal phase mode (NP) requires the employment of a polar column together with a gradient whose polarity increases in the course of the run. In a proper NP system retention increases with sample polarity. Gradient elution in the reversed phase mode (RP) requires the use of a non-polar column together with a gradient whose polarity decreases in the course of the run. In an RP system retention decreases with increasing sample polarity. The opposite is the case for NP systems. With styrene-ethyl methacrylate copolymers it was demonstrated that mixed samples can be eluted in sequences increasing as well as decreasing in ethyl methacrylate content [16].

Mobile Phases. For the selection of mobile phases the following criteria must be observed: solubility of the sample, controlled interactions with the solute and the stationary phase, suitability for a specific detector, chemical stability, miscibility of mobile phases as a precondition for gradient techniques, solubility of buffers and modifiers, viscosity, and environmental compatibility. The classification of mobile phases in interaction chromatography usually is done with respect to the solvent strength or the elution power, and the so called "eluotropic series" compares different solvents in the order of increasing solvent strength ε^0. The eluotropic series must be determined experimentally for each stationary phase by determining the retention time as a function of the composition of the eluent. For stationary phases of different polarity, the eluotropic series will be completely different. Thus, for silica gel as the stationary phase, n-heptane exhibits the lowest elution power ($\varepsilon^0=0$), whereas i-propanol is a strong eluent ($\varepsilon^0 = 0.82$). In reversed-phase chromatography, water is the weakest eluent, and n-heptane exhibits a high elution power.

Detectors. Once the chromatographic separation on the column has been conducted, the composition of the eluent must be determined using a detector. In all HPLC detectors, the eluent flows through a measuring cell, where the change of a physical or chemical property with elution time is detected. The most important parameter of the detector is sensitivity, which is influenced by the noise and baseline drift, the absolute detection limit of the detector, the linearity, the detector volume (band broadening), and the effects of pressure, temperature and flow (pulsation, gas bubbles).

In gradient elution, detection requires quantitative measurement of the sample components in an eluent whose composition and, hence, physical properties alter in the course of the analysis. The detection problem in gradient elution can be solved by either using a selective detector sensitive to a property of only the solute or stripping off the solvent with subsequent measurement of non-volatile residues.

The current status and further prospects of HPLC detectors have been dealt with in detail in a number of reviews [17–19]. In brief, universal and selective detectors are used in liquid chromatography. With universal detectors, the solute concentration in the eluate may be detected regardless of the chemical structure of the solute. The most commonly used universal detector is the differential refractometer, which measures changes of the refractive index of the polymer solution as compared to the refractive index of the pure solvent. This detector can be used only in isocratic elution. The change in composition of the eluent in gradient elution and the corresponding change in the refractive index of the eluent causes a significant baseline drift. In gradient elution, the evaporative light-scattering detector (ELSD) can be used as a universal detector despite response linearity problems. Any non-volatile solute may be detected. If the eluent contains buffer salts, then these should be volatile. The eluate from the HPLC column is directed into a nebulizer where a gas supply atomizes the liquid stream. The droplets are sprayed into the evaporator where the solvent is vaporized. The resulting aerosol is then measured via light scattering.

The most commonly used selective detectors are photometers for UV and visible light. The accessible wavelength range is 190–800 nm, but depending on the light absorption properties of the solute and the eluent specific wavelengths are usually selected. The photo diode array detector is capable of monitoring the complete UV spectrum during the chromatographic run. Addi-

tional information on the structure of the sample components
may be obtained. Even more detailed structural information may
be obtained by using an IR detector. However, absorption bands
of the eluent may disturb the detection. In a new approach, a
deposition technique is applied, in which the eluent is removed
from the sample enabling the complete IR spectra of the sample
components to be measured [20, 21]. Of particular interest for
polymer analysis are molar mass sensitive detectors, i.e. viscosity, and light scattering detectors [22–24].

5.3 Data Acquisition and Processing

The aim of liquid adsorption chromatography is the separation
of samples with respect to chemical composition. As a result,
chromatographic peaks are obtained which represent chemically
different fractions of the sample. The desired information for
each fraction is its relative concentration and chemical composition.

The chromatographic peak, obtained as the result of the separation, bears a qualitative and a quantitative information. The qualitative information is obtained from the retention time of a particular component and is related to the chemical composition of this
component. In copolymer analysis the retention time is correlated
with chemical composition via a corresponding calibration procedure, where samples with known chemical composition are
investigated with regard to their elution time behaviour.

The quantitative information relates to concentration and
can be obtained from the area or the height of a particular elution peak. Calibration curves peak area vs concentration can
be constructed using calibration standards, which can be of
two types: external standards or internal standards. With
external standards multiple concentrations of the standards
are injected, areas are measured, and a calibration curve is
plotted. Unknowns are then run, and areas are calculated and
compared to the calibration curves to determine amounts of
each component present. With internal standards, known
amounts of an internal standard are added to each known concentration of standard compounds and area or peak height
response factors relative to those of the internal standard are
calculated. When unknowns are run, the same amount of internal standard is added to the unknown sample, and relative
areas or heights are calculated based on the response factor of
the internal standard from the calibration curves.

Different from adsorption chromatography of low molar mass
compounds and oligomers, in adsorption chromatography of

polymers there are several pitfalls which can cause unnoticed loss of sample, incomplete elution, or imperfect separations. Effects of this kind can not only lead to incorrect quantitative results but also may affect qualitative conclusions. Since these effects disturb the linear dependence of peak area on sample size, careful quantification using suitable calibrants is recommended.

Quantification in polymer HPLC is difficult because, owing to the distribution in size and composition of the samples, broad peaks are obtained frequently. Proper baseline control is of key importance, otherwise integration can yield incorrect results. Special care is required with non-linear baselines due to different gradient programs. Here, point-to-point reading of the difference between chromatogram and blank-gradient baseline is necessary.

In some cases incomplete retention of a sample can cause excluded elution, elution close to the solvent peak, or elution with the sample solvent. In particular, the last case is the worst one, because a part of the sample is swept through the column by the solvent plug. This portion then may easily elute unnoticed and, accordingly, is not accounted for in the quantification procedure. The goal in any case must be proper retention of the whole sample. On the other hand, care must be taken that the solvent strength of the eluent is high enough to elute all sample components from the column. Since often the chromatographic behaviour of a sample cannot be predicted, it is useful to flash the column after the separation step with a good solvent. The elution of an additional sample fraction under these conditions indicates that there was any fraction of the sample left undesorbed on the column. As has been discussed, in some cases adsorption chromatography involves precipitation steps. In a system of that kind, elution occurs in an eluent whose composition cannot guarantee a stable solution. The eluting polymer may be segregated in fine droplets of a gel phase which, in an optical detector, would cause a signal which may be enhanced due to turbidity effects.

The quantitative evaluation of copolymer chromatograms with regard to chemical composition requires the knowledge of the influence of polymer composition on elution time, the influence of molar mass on elution time, and the influence of composition on detector signal. This knowledge can be gained by calibration with a series of samples graded in composition. Since composition and molar mass effects on elution time are interrelated, both effects are at best evaluated simultaneously. This can be done by SEC fractionation of a calibrant and adsorption chromatography of the fractions. Through the use of universal calibration, the molar mass of each fraction can be calculated from sample composition and SEC elution volume [2, 25].

5.4 Analysis of Oligomers

The analysis of oligomers is an important application of polymer LAC. Depending on the complexity of the sample and the molar mass range, a complete separation into individual oligomers can be achieved. This is rather straightforward for chemically homogeneous oligomer samples. Since these types of samples are distributed only with respect to chain length, a uniform homologous series is expected to be obtained.

5.4.1 Separation of Octylphenol Ethoxylate by Isocratic Elution

Aim

Alkylphenol and fatty alcohol ethoxylates are important technical products for use as surfactants. They are produced by anionic or cationic ring opening polymerization of ethylene oxide in the presence of the corresponding phenol or alcohol. In the ideal case alkylphenyloxy- or alkoxy-terminated polyethylene oxides (PEO) are formed, which are homogeneous with respect to the end group and distributed with respect to chain length. The chain length distribution shall be determined by conducting an oligomer separation.

Materials

Polymer. Technical octylphenyloxy-terminated polyethylene oxide with an average degree of polymerization of six.

Equipment

Chromatographic System. Waters LC Module 1 for gradient elution equipped with Millennium 2010 chromatography manager.

Columns. Nucleosil RP-18 of Macherey-Nagel, 5 µm average particle size and 100 Å average pore diameter. Column size was 250x4 mm I.D.

Mobile Phase. Mixtures of acetonitrile and water, all solvents are HPLC grade.

Detectors. Built-in Waters tunable UV detector.

Column Temperature. 25 °C.

Sample Concentration. 5 mg/mL. The samples are dissolved in acetonitrile.

Injection Volume. 10 µL.

The quality of the separation is influenced by the surface activity
of the stationary phase and the composition of the mobile phase.
Accordingly, for the present sample a number of test runs should
be conducted using different stationary phases and eluents. However, successful separations can be carried out on bonded silica
gel. For this type of reversed-phase HPLC, aqueous mobile phases such as water-methanol, water-THF, and water-acetonitrile
can be used [26–29].

    In the present case a reversed-phase column RP-18 is used, for
simplicity isocratic elution with water-acetonitrile is applied. For
optimizing the separation, mobile phase compositions of water-
acetonitrile 30:70, 40:60 and 50:50% by volume are recommended.

An optimum separation of the octylphenyloxy-terminated PEO
is obtained using a mobile phase of water-acetonitrile 50:50% by
volume. In this case, an appropriate flow rate is 1 mL/min. Since
the oligomers have aromatic end groups, they can be detected
readily with a UV detector. A maximum peak intensity is
obtained using a wavelength of 220 nm.

    The separation of the sample into oligomers is shown in Fig.
5.6. In contrast to what is expected for a chemically uniform sample, a complex chromatogram is obtained, showing three different oligomer series. The major oligomer population, series 1, in
the retention time range of 17–35 min corresponds to the
octylphenol ethoxylate, each peak separated by one degree of
polymerization. A second oligomer population, series 2, of lower
intensity is obtained at lower retention time between 6 and 9
min. Since separation in this case is due to hydrophobic interactions, this oligomer series must be more polar than the major
oligomer series. This is in agreement with the matrix-assisted
laser desorption/ionization (MALDI) mass spectrometric analysis, which shows that series 2 is caused by butylphenol ethoxylates. Ostensibly, butylphenol was present in the reaction mixture
as an impurity of the technical octylphenol, which was used as
the starting reagent in the ethoxylation reaction.

    Oligomer series 2 is followed by a third series of very low inten-
sity, which elutes in the range of 9–14 min. Considering the
results of oligomer series 2, this oligomer series 3 is assigned to
alkylphenol ethoxylates, where the alkyl group may be pentyl or
hexyl. Because of the very low concentration of this fraction,
identification by MALDI-MS could not be carried out.

    All three oligomer series are separated in the order of decreas-
ing degree of polymerization, which is also common for SEC separations. However, in the present case separation is directed by
hydrophobic interactions. Since oligomers with a long ethylene
oxide chain (high degree of polymerization) are less hydrophobic

*Preparatory
Investigations*

Separations

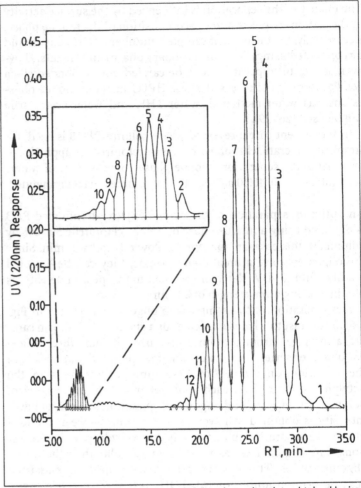

Fig. 5.6. Chromatogram of a technical octylphenyloxy ethoxylate obtained by isocratic elution stationary phase: Nucleosil RP-18; mobile phase: water-acetonitrile 50:50% by volume; numbers indicate degree of polymerization.

than the lower oligomers, they elute first, followed by oligomers of lower degree of polymerization.

Evaluation

Since the chromatographic run yields well-separated single oligomers for oligomer series 1, these oligomers can be subjected to MALDI-MS for identification. By this procedure, the degree of polymerization (n) and, accordingly, the molar mass M can be determined for each oligomer. The assignment of the peaks to a certain n (see Fig. 5.6) agrees perfectly with the assignment of the MALDI spectrum and the SFC chromatogram of series 1 (see Fig. 5.7).

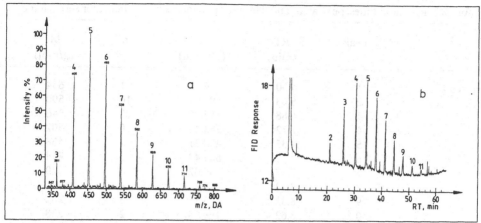

Fig. 5.7. MALDI spectrum (a) and SFC chromatogram (b) of oligomer series 1, assignment represents degree of polymerization (n)

The detector response of the oligomers results solely from the UV absorption properties of the octylphenol end group. Since each oligomer contains exactly one of these end groups, the peak area in the chromatogram is directly related to concentration. This, of course, holds true also for oligomer series 2 and 3.

The expanded part of Fig. 5.6 represents oligomer series 2, which was found to be butylphenol ethoxylate. The shape of this oligomer series is very similar to series 1. The addition of ethylene oxide to butylphenol and octylphenol can be assumed to occur with the same reaction rate, and accordingly, similar oligomer series can be expected to be formed. Based on this consideration, the assignment of the oligomer peaks in series 2 to a degree of polymerization is carried out. The analysis of the chromatogram (Fig. 5.6) with respect to oligomer series 1 and 2 is given in Table 5.3.

From n and the mass of the end group, the molar mass (M) of each oligomer can be calculated. The average molar masses M_w and M_n of the oligomer series can then be calculated from M and the peak areas A_i, which correspond to molar concentration, since UV detection is only sensitive to the aromatic end group. Assuming that the UV response of the butyl- and octylphenol end groups are rather similar, the molar ratio of oligomer series 1 and 2 can be calculated by summing the corresponding peak areas.

$$M_i \text{ (series 1)} = 206 + 44\,n \qquad (5.1)$$

$$M_i \text{ (series 2)} = 150 + 44\,n \qquad (5.2)$$

$$\text{amount series } 1 = \Sigma\, A_i^{(1)} / \Sigma\, A_i^{(1)} + \Sigma\, A_i^{(2)} \qquad (5.3)$$

Table 5.3. Retention times, peak areas and degree of polymerization of oligomer series 1 and 2

	Peak	RT (min)	A_i (10^3 a.u.)	n	M_i (g/mol)
series 2	1	6.50	76.99	11	634
	2	6.68	95.04	10	590
	3	6.88	184.00	9	546
	4	7.12	283.44	8	502
	5	7.35	430.21	7	458
	6	7.62	541.43	6	414
	7	7.87	699.02	5	370
	8	8.12	608.14	4	326
	9	8.32	455.15	3	282
	10	8.62	169.58	2	238
series 1	1	17.52	46.79	15	866
	2	18.10	140.69	14	822
	3	18.72	326.92	13	778
	4	19.38	737.34	12	734
	5	20.12	1498.45	11	690
	6	20.92	2770.70	10	646
	7	21.80	4730.70	9	602
	8	22.78	7525.36	8	558
	9	23.87	10879.96	7	514
	10	25.08	14198.56	6	470
	11	26.17	16639.25	5	426
	12	27.32	16048.60	4	382
	13	28.45	11272.80	3	338
	14	30.05	5019.97	2	294
	15	32.42	743.36	1	250

$$\text{amount series 2} = \Sigma\, A_i^{(2)}/\Sigma\, A_i^{(1)} + \Sigma\, A_i^{(2)} \qquad (5.4)$$

$$M_n = \Sigma\, A_i M_i / \Sigma\, A_i \qquad (5.5)$$

$$M_w = \Sigma\, A_i M_i^2 / \Sigma\, A_i M_i \qquad (5.6)$$

where $A_i^{(1)}$ and $A_i^{(2)}$ are the molar fractions of the i-th component of series 1 and 2, respectively.

For the present sample the following composition and molar masses are calculated:

	Molar Conc. (mol%)	M_w (g/mol)	M_n (g/mol)
series 1	96.3	474	451
series 2	3.7	423	406

5.4.2 Separation of Oligostyrene by Gradient Elution

Oligomer mixtures of synthetic polymers may be heterogeneous with respect to chemical composition and molar mass. As was shown in Section 5.4.1, high resolution LAC is capable of providing information on both types of heterogeneity. For oligomer mixtures of homopolymers which do not exhibit a functional heterogeneity, only a molar mass distribution is expected.   Aim

However, in an early study of Eisenbeiss et al. [30] it was shown that oligostyrene is separated into more than one homologous series. A possible reason for this behaviour may be the presence of tactic isomers. In the present experiment, a number of oligostyrenes will be compared with respect to their chromatographic behaviour.

Polymer. Oligostyrene calibration standards of different molar masses.   Materials

Chromatographic System. Waters LC Module 1 for gradient elution equipped with Millennium 2010 chromatography manager.   Equipment

Columns. Nucleosil RP-18 of Macherey-Nagel, 5 µm average particle size and 100 Å average pore diameter. Column size was 250x4 mm I.D.

Mobile Phase. Acetonitrile HPLC grade.

Detectors. Built-in Waters tunable UV detector, measurements are carried out at 260 nm.

Column Temperature. 25 °C.

Sample Concentration. 5 mg/mL. The samples are dissolved in acetonitrile.

Injection Volume. 50 µL.

The separation of oligostyrenes on a reversed-phase column can be carried out using different mobile phases. With THF as the mobile phase, conventional SEC separation is obtained. The adsorption mode is operating when acetonitrile is used as the eluent. In this case separation is accomplished predominantly by enthalpic interactions and the oligomers elute from the column in the order of increasing molar masses. Since   Preparatory Investigations there is no sense in further modifying the eluent, a flow rate gradient is used to optimize the separation. With increasing molar mass the oligomer peaks become very broad. Narrower peaks are obtained, when the flow rate is increased in the course of the separation. A number of initial tests varying the flow rate are used for optimization.

Sufficiently well separated and narrow peaks are obtained by using the following flow rate program: A linear flow rate increase is used in the present case.

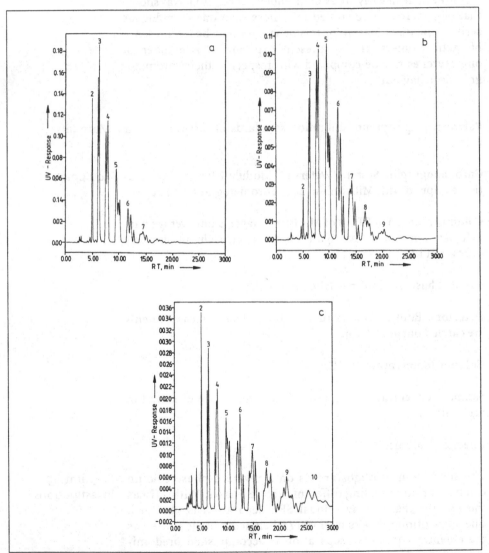

Fig. 5.8. Chromatograms of oligostyrenes with molar masses of 530 **(a)** 690 **(b)** and 1010 g/mol **(c)**; stationary phase: Nucleosil RP-18; mobile phase: acetonitrile; numbers indicate degree of polymerization

Time (min)	Flow Rate (mL/min)
0	1.0
8	1.0
20	2.5
25	2.5
30	1.0

The chromatograms of three different oligostyrenes are summarized in Fig. 5.8. In all cases separations into different oligomers are obtained. The degree of polymerization n, which was determined by comparison with model oligomers, is indicated at the corresponding elution peaks. Depending on n, more or less complex substructures for each oligomer peak are obtained. For n=2 a substructure is not observed and the oligomers are eluted in one single peak.

Oligomers of n=3 are eluted in two peaks, n=4 yields two to three peaks, and higher oligomers show even more complex substructures. For n=9 a minimum of seven subpeaks is obtained.

The retention times for the oligomers of the different samples are in good agreement, indicating that the same chemical structures are present.

Since all three samples are SEC calibration standards, chemical uniformity can be assumed. The only possible explanation for the peak substructures is the presence of different isomeric structures. Depending on the preparation procedure of the samples, isotactic and syndiotactic sequences can be formed in different ratios. Possible architectures for the first oligomers are:

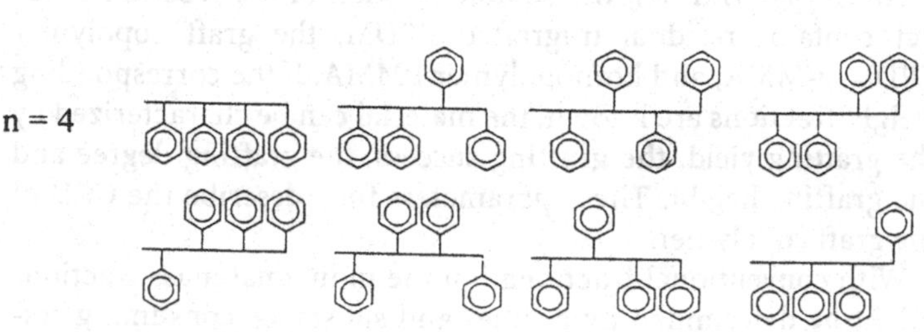

Separations *(margin)*

Evaluation *(margin)*

n = 2

n = 3

n = 4

An identification of the different isomers is difficult. One way would be a preparative separation and identification of the fractions by NMR spectroscopy. Much more elegant would be an on-line HPLC/NMR experiment, which is assumed to become available in the near future [31, 32].

5.5 Fractionation of Copolymers

Copolymers are complex macromolecular systems which are formed when two or more monomers of different chemical structures react in a polymerization process. As a result, products are obtained which, in addition to the MMD, are characterized by a certain distribution in chemical composition (CCD). CCD describes the sequence distribution of the different monomer units along the polymer chain.

The complex molar mass-chemical composition distribution of copolymers requires separation in more than one direction. The classical approach is based on the dependence of copolymer solubility on composition and chain length. A solvent/nonsolvent combination fractionating solely by chemical composition would be appropriate for determining the CCD of the copolymer. This technique can be used also in column liquid chromatography. HPPLC was used by Glöckner to fractionate copolymers with respect to CCD [2].

5.5.1 Analysis of Graft Copolymers of Ethylene-Propene-Diene Rubber and Methyl Methacrylate [15]

Aim

Graft copolymers are interesting polymer materials for application as compatibilizer in polymer blends, as emulsifying agent or dispersant. With respect to molecular heterogeneity, graft copolymers are characterized by a distribution in chemical composition in addition to the molar mass distribution. When ethylene-propene-diene rubber (EPDM) is grafted with methyl methacrylate (MMA), the soluble fraction of the reaction product contains residual ungrafted EPDM, the graft copolymer EPDM-g-MMA, and homopolymer PMMA. If the corresponding weight fractions are known, the material can be characterized by the grafting yield, the grafting success, the grafting degree and the grafting height. These parameters fully describe the CCD of the graft copolymer.

With conventional fractionation the individual mass fractions might be determined by a time- and substance-consuming solvent extraction or solution/precipitation procedure. The present experiment is aimed at using fast and reliable chromatographic separation for the analysis of graft copolymers.

Calibration Standards. Narrow-disperse PMMA in the molar mass range of 10^3–10^6 g/mol.

Materials

Polymers. The EPDM-g-MMA are laboratory products of Röhm GmbH, Darmstadt, Germany. They are prepared by radical grafting of MMA onto EPDM using tert-butyl peroxy-2-ethylhexanoate as the initiator.

Chromatographic System. Commercial HP 1090 chromatograph with work station and built-in diode array detector.

Equipment

Columns. HPLC: Nucleosil 5 CN (Knauer, Germany), 5 µm average particle size and 100 Å average pore size, with a column size of 60x4 mm I.D.; SEC: two TSK GMH 6 columns (Toyo Soda, Japan), 10 µm average particle size, with column sizes of 600x7.5 mm I.D.

Mobile Phase. HPLC: isooctane-THF, starting from 99:1% by volume, being held constant for 1 min and then being changed linearly to 100% THF within 6 min; SEC: THF.

Detectors. HPLC: ELSD mass detector model 750 (Zinsser-ACS), SEC: ERMA 7510 differential refractometer (ERC, Germany) and UV detector BT 3030 (Biotronik, Germany).

Column Temperature. 40 °C (HPLC).

Sample Concentration. 1 mg/mL (SEC) and 0.1–5 mg/mL (HPLC).

Injection Volume. 200 µL (SEC) and 5 µL (HPLC).

First information on the chemical structure of the graft copolymer can be obtained by SEC. Since EPDM does not absorb light at 239 nm, whereas PMMA does, some information on the MMD and chemical heterogeneity can be gained by using dual RI/UV detection. With the use of PMMA for calibration, in a first approximation the overall MMD is obtained from the RI detector tracing (see Fig. 5.9). By UV detection, only the MMA-containing species are detected. Accordingly, the UV tracing indicates homopolymer PMMA and the side chains of the graft copolymer. The low-molar-mass peak of the bimodal distribution exhibits a high UV activity.

Preparatory Investigations

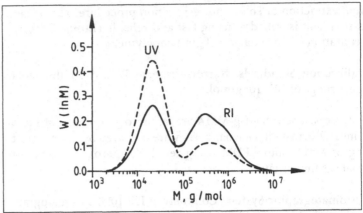

Fig. 5.9. Molar mass distribution of the reaction product of grafting MMA onto EPDM from SEC analysis; stationary phase: styragel; mobile phase: THF; detectors: UV and RI detectors (Reprinted from Ref. [15] with permission of Hüthig & Wepf Publishers, Switzerland)

Assumingly, this peak is due to PMMA homopolymer, whereas the higher molar mass peak contains the EPDM precursor and the grafted product. The ungrafted EPDM and the graft copolymer give overlapping elution profiles, so that their amounts cannot be determined from these distributions alone.

Separations

The constituents of the graft product under investigation can be expected to differ greatly in polarity: EPDM does not contain any polar groups that could interact with each other or with polar groups on the surface of the packing. In addition, EPDM is soluble in isooctane. PMMA is relatively polar and will precipitate in isooctane. Graft copolymers should reflect increasing polarity with increasing MMA content.

In order to separate the reaction product with respect to polarity and solubility, solutions in THF are injected on nitrile-modified silica gel. The starting eluent is isooctane-THF 99:1% by volume. Since the stationary phase is polar, the retention increases with increasing polarity of the solute molecules. Accordingly, the EPDM-rich fraction should be eluted first, followed by fractions increasing in MMA content. In addition, PMMA is not soluble in the starting eluent and will precipitate in the column, and will leave the column last.

In order to avoid any mobile phase influence on the detector signal, an evaporative light scattering detector (ELSD) was used for the HPLC experiments. The ELSD tracing for the gradient elution of the product of the grafting reaction is shown in Fig. 5.10. Although a solvent gradient of isooctane-THF is used, it is not detected and the baseline is perfectly straight.

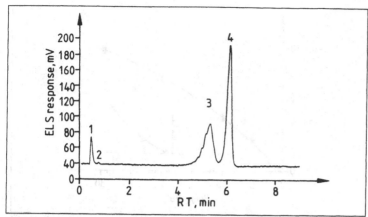

Fig. 5.10. ELS detector tracing for the gradient elution of the grafting product; stationary phase: Nucleosil 5 CN; mobile phase: isooctane-THF (Reprinted from Ref. [15] with permission of Hüthig & Wepf Publishers, Switzerland)

There are three elution peaks visible that can be assigned to the sample constituents EPDM, PMMA and EPDM-g-MMA. Based on the assumption of the elution order and comparison with the starting material, the first peak can be assigned to ungrafted EPDM. This component is soluble in the starting mobile phase and does not undergo interactions with the polar stationary phase. Accordingly, it is eluted quickly and leaves the column first. The third peak is significantly retained, indicating that it contains polar MMA units. Based on comparison with a model compound, the fourth peak can be assigned to PMMA homopolymer. As was expected, it is retained most strongly on the stationary phase. Following the assignment of peaks 1 and 4, peak 3 can be assigned to EPDM-g-MMA, which is eluted according to the amount of MMA grafts. The small peak 2 is assumed to be from a solvent plug, which can be minimized using minimum injection volumes for sample introduction.

Evaluation

The quantitative analysis of the chromatogram yields the weight fractions of ungrafted EPDM and homopolymer PMMA. Since the ELSD signal is a function of refractive index, average size and size distribution of the scattering particles, a calibration of the detector with respect to EPDM and PMMA must be carried out [33–35]. The PMMA response curve, given in Fig. 5.11, shows a non-linear behaviour, the mathematical treatment of this type of calibration curve, however, is not a problem using appropriate data processing units.

Knowing the weight fractions of EPDM and PMMA, one can calculate the weight fraction of EPDM-g-MMA. From these data then, all grafting parameters are available and can be calculated using the following formulae:

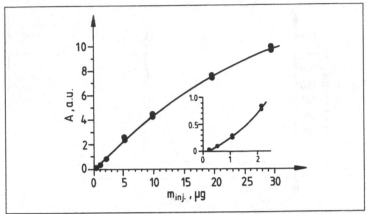

Fig. 5.11. ELSD response curve for PMMA: peak area A vs injected mass of sample. The inset gives a magnified representation for small sample amounts (Reprinted from Ref. [15] with permission of Hüthig & Wepf Publishers, Switzerland)

$$\text{grafting yield } a = m_1/m_2 \qquad\qquad (5.7)$$

$$\text{grafting success } e = m_3/m_4 \qquad\qquad (5.8)$$

$$\text{grafting degree } g = m_1/m_4 \qquad\qquad (5.9)$$

$$\text{grafting height } h = m_1/m_3 \qquad\qquad (5.10)$$

where m_1 is the mass of MMA grafted onto EPDM, m_2 is the total mass of consumed MMA, m_3 is the mass of grafted EPDM, and m_4 is the total mass of EPDM. The quantitative results for a typical sample are summarized in Table 5.4.

More detailed information on the distribution of molar mass and chemical composition can be obtained carrying out a SEC-HPLC cross fractionation. Figure 5.12 gives a qualitative representation of the MMD-CCD for the sample under investigation. It contains the HPLC elution profiles of successive SEC fractions as detected by ELSD. The fractions are collected from a number of identical SEC runs, evaporated to dryness and redissolved for the HPLC experiments. Fractions of high molar mass show broad peaks between 4 and 6 min corresponding to peak 3 in Fig. 5.10. The range of molar mass and the elution just before the PMMA fraction (peak 4) prove that these signals represent the graft copolymer.

The main features of Fig. 5.12 offer a detailed view on the correlation of chemical composition with molar mass. By SEC prefractionation, it is apparently possible to avoid any overlap between the elution profiles of graft copolymer and PMMA; a

Table 5.4. Weight fractions w of produced homopolymer PMMA and ungrafted EPDM as well as grafting parameters for a sample with the initial weight fractions of rubber and monomer: w(EPDM)=0.333, w(MMA)=0.667

Parameter	SEC UV (239 nm)	HPLC ELSD
w(PMMA)	0.41	0.41
w(EPDM)		0.064
a (Eq. 5.7)	0.38	0.39
e (Eq. 5.8)		0.81
g (Eq. 5.9)	0.76	0.78
h (Eq. 5.10)		0.96

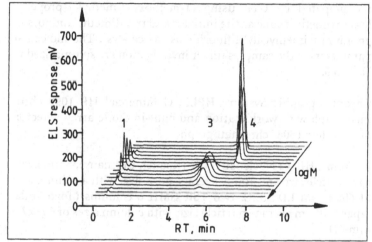

Aim

Fig. 5.12. Superposition of the gradient elugrams as detected with ELSD for the SEC fractions of the sample (Reprinted from Ref. [15] with permission of Hüthig & Wepf Publishers, Switzerland)

close inspection of this figure shows that the graft components of highest molar mass and the PMMA species of lowest molar mass are responsible for the overlap observed in Fig. 5.10.

5.5.2 Analysis of Copolymers of Decyl and Methyl Methacrylate [36]

Group transfer polymerization (GTP) is one of the techniques to polymerize methacrylates by a living mechanism yielding narrow MMD [37, 38]. In addition to random copolymers, block copolymers can be prepared, if the monomers are added successively [39]. Owing to side reactions upon addition of the second monomer, some of these precursor molecules fail to continue the chain growth and are left as a homopolymer portion contaminating the resulting block copolymer.

The investigation of the behaviour of random and block copolymers in gradient HPLC shall be compared and quantitative information on homopolymer contaminations and the copolymer composition shall be obtained.

Materials

Calibration Standards. Narrow-disperse PMMA in the molar mass range of 10^3–10^6 g/mol.

Polymers. The copolymers of decyl and methyl methacrylate are laboratory products of Röhm GmbH, Darmstadt, Germany. They are prepared by GTP using (1-methoxy-2-methyl-1-propenyloxy)-trimethylsilane as the initiator and tris(dimethylamino)sulphonium trimethylsilyldifluoride as the catalyst. The molecular parameters of the samples under investigation are summarized in Table 5.5.

Equipment

Chromatographic System. HPLC: Commercial HP 1090 chromatograph with work station and built-in diode array detector, SEC: Waters 150 C chromatograph.

Columns. HPLC: Nucleosil 5 CN (Knauer, Germany), 5 µm average particle size and 100 Å average pore size, with a column size of 60x4 mm I.D.; SEC: two TSK GMH 6 columns (Toyo Soda, Japan), 10 µm average particle size, with column sizes of 600x7.5 mm I.D.

Mobile Phase. HPLC: isooctane-THF, starting from 99:1% by volume, being held constant for 1 min and then being changed linearly to isooctane-THF 19:81% by volume within 8 min; SEC: THF.

Table 5.5. Molecular parameters of the copolymer samples

	Sample	DMA/MMA (molar ratio)	M_w(SEC) (g/mol)	M_n(SEC) (g/mol)
1.	P(DMA-st-MMA)	0.75/0.25	125.000	110.000
2.	P(DMA-b-MMA)	0.75/0.25	97.000	88.000
3.	P(MMA-b-DMA)	0.50/0.50	135.000	101.000

Detectors. HPLC: ELSD mass detector model 750 (Zinsser-ACS), SEC: Waters differential refractometer (Waters, Germany) and UV detector BT 3030 (Biotronik, Germany).

Column Temperature. 40 °C (HPLC).

Sample Concentration. 1 mg/mL (SEC) and 5 mg/mL (HPLC).

Injection Volume. 300 μL (SEC) and 5 μL (HPLC).

Decyl (DMA) and methyl methacrylate (MMA) are polymerized   **Preparatory** by GTP to obtain random and block copolymers of narrow   **Investigations** MMD. Since block copolymers are suspected to contain portions of homopolymer, depending on the preparation procedure, different homopolymers can be encountered. If DMA constitutes the first block, PDMA precursor may be left; if MMA is polymerized first, PMMA may be present in the reaction mixture.

The MMD of the reaction products can easily be determined by SEC. Figure 5.13 shows the distribution curves for samples 1–3. The polydispersities are low as expected from GTP. The MMDs are calculated using a PMMA calibration curve; the calibration curves for both homopolymers PMMA and PDMA proved to be identical within the limits of experimental error.

Comparison of the chromatograms reveals no difference between the random (A) and the block copolymer (B). Neither chromatogram gives any indication of homopolymer formation. The formation of homopolymer is indicated in (C) with the appearance of a bimodal distribution. Here the lower molar mass peak can be assigned to the homopolymer fraction.

The separation of the copolymers with respect to chemical com-   **Separations** position is carried out by gradient HPLC. Similar to Section 5.5.1,

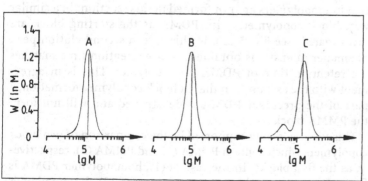

Fig. 5.13. Molar mass distributions of samples 1 (A) 2 (B) and 3 (C) from SEC; stationary phase: Styragel; mobile phase: THF (Reprinted from Ref. [36] with permission of Hüthig & Wepf Publishers, Switzerland)

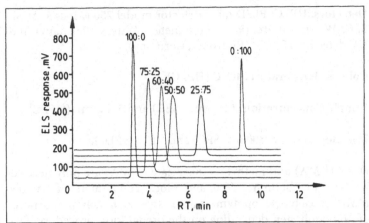

Fig. 5.14. Gradient HPLC chromatograms of random copolymers of DMA and MMA with various compositions given as mol.% DMA/MMA; stationary phase: Nucleosil 5 CN; mobile phase: isooctane-THF (Reprinted from Ref. [36] with permission of Hüthig & Wepf Publishers, Switzerland)

the starting eluent composition is a non-solvent for MMA-rich fractions and, therefore, the HPPLC mode is in operation. In practice it is often impossible to define clearly to what extent adsorption and solubility (precipitation) contribute to retention. From the present work it is obvious that the relative contributions vary with the chemical composition of the solute. PDMA is soluble in isooctane and therefore retained only by adsorption, whereas PMMA and most copolymers of MMA and DMA are not soluble in isooctane so that solubility is a major parameter.

Figure 5.14 shows the ELSD tracings for the homopolymers PMMA and PDMA and four random copolymers of different compositions. As expected, the retention time increases with increasing content of the more polar MMA. The chromatograms show a single symmetric elution peak, indicating that homopolymers are not formed in this reaction. In a similar way, block copolymers with PDMA as the starting block are investigated (see Fig. 5.15). In this case, a second elution peak of smaller intensity is obtained, whose retention time matches the retention time of PDMA homopolymer. This is in agreement with the assumption that in block copolymer formation, a part of the precursor PDMA is deactivated and will not form the PMMA block.

This finding is fully supported by the comparison of two block copolymers which contain PDMA (1) and PMMA (2), respectively, as the first blocks. In the case of (1), homopolymer PDMA is formed as a by-product, while in case of (2), homopolymer PMMA is obtained (see Fig. 5.16).

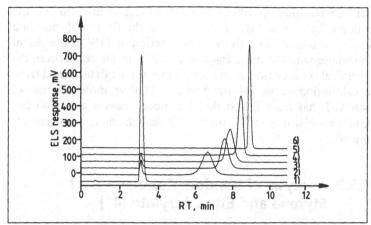

Fig. 5.15. Gradient HPLC chromatograms of block copolymers of DMA and MMA with various compositions given as mol.% DMA/MMA: 100:0 (1) 75:25 (2) 60:40 (3) 50:50 (4) 25:75 (5) 0:100 (6); stationary phase: Nucleosil 5 CN; mobile phase: isooctane-THF (Reprinted from Ref. [36] with permission of Hüthig & Wepf Publishers, Switzerland)

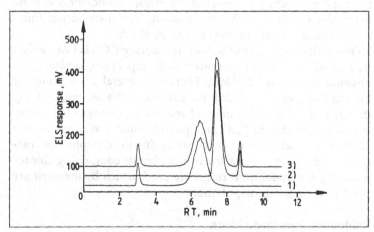

Fig. 5.16. Gradient HPLC chromatograms of block copolymers P(DMA-b-MMA) (1) and P(MMA-b-DMA) (2) and a mixture of these polymers (3); stationary phase: Nucleosil 5 CN; mobile phase: isooctane-THF (Reprinted from Ref. [36] with permission of Hüthig & Wepf Publishers, Switzerland)

Quantitative analysis of the chromatograms yield the weight fractions of the sample components, compare Section 5.5.1. The determination of the chemical composition of the copolymer fractions is possible in two ways: The most convenient way is the calculation of chemical composition from the monomer feed, the polymerization yield, the molar mass of the copolymer and the portions of homopolymer, which may be determined from the

Evaluation

HPLC separation. Another possiblity is the determination via the volume fraction of THF in the eluent at the time of elution of a certain fraction. Since the volume fraction of THF in the eluent correlates with the mole fraction of MMA in the copolymer, the chemical composition of the copolymer can be determined from a calibration curve volume fraction THF vs mole fraction of MMA. It has been found that for molar masses above 60 000 g/mol a molar mass effect on the elution behaviour is not operating [40].

5.5.3 Analysis of Random Copolymers of Styrene and Ethyl Acrylate [41]

Aim

Random copolymers have a molar mass distribution and a distribution in chemical composition. The accurate determination of CCD is important for the characterization of copolymers, because a number of product properties are influenced by this parameter. For copolymers comprising an aromatic and an aliphatic monomer, the overall chemical composition can be determined easily by UV spectroscopy. The distribution function, however, is not available by this method.

One of the major techniques for measuring CCD is LAC, which was found to separate copolymers with respect to chemical composition regardless of MMD. There are several applications of LAC in this area [12, 42–45], the essence of the technique being the use of a combination of good and poor solvents as the eluent. It is assumed that in LAC precipitation does not occur on the packing, and separation is exclusively due to adsorptive interactions. In the present experiment two eluents comprising chloroform/ethanol in different ratios are used, which by themself are good solvents for the copolymers.

Materials

Calibration Standards. None.

Polymers. The styrene-ethyl acrylate random copolymers (S-EA) were prepared by radical polymerization in benzene using AIBN as the initiator. The copolymerization was carried out only to a low degree of conversion in order to keep the CCD as low as possible. The molar mass averages of the copolymers measured by SEC were between 1.0 and 3.0x10^5 g/mol for M_w. The overall chemical composition was determined by UV spectroscopy (seeTable 5.6).

Equipment

Chromatographic System. Trirotar-VI high-performance liquid chromatograph (Jasco, Japan) with a model TU-300 column oven.

Table 5.6. Overall chemical composition of the copolymer samples

Sample	Styrene/Ethyl Acrylate (molar ratio)
1	0.686/0.314
2	0.526/0.474
3	0.367/0.633
4	0.207/0.793

Columns. Silica gel (Nomura, Japan), 5 µm average particle size and 30 Å average pore size, with a column size of 50x4.6 mm I.D.

Mobile Phase. Mixtures of chloroform and ethanol, from commercial chloroform the ethanol stabilizer was removed prior to use. Linear gradient elution was performed as follows: The initial mobile phase A was chloroform-ethanol 99:1% by volume, the composition of the final mobile phase B was chloroform-ethanol 93:7% by volume. The composition of the mobile phase was changed from 100% A to 100% B in 30 min linearly, and then the mobile phase was returned to 100% A in another 10 min.

Detectors. UV spectrometer Uvidec-610C (Jasco, Japan).

Column Temperature. variable 40–70 °C.

Sample Concentration. 1 mg/mL, all samples are dissolved in the mobile phase.

Injection Volume. 25 µL.

In a previous study it was reported that styrene-methyl methacrylate copolymers tend to be retained on silica gel when chloroform was used as the mobile phase [46]. They were eluted when ethanol was added to the mobile phase. The copolymers having more of the methyl methacrylate component required a higher ethanol content in the mobile phase and a lower column temperature for elution.

In order to examine the elution behaviour of the present S-EA samples with respect to mobile phase composition, isocratic experiments at 50 °C column temperature are conducted. Samples 2 and 3 are measured in mobile phases containing different amounts of ethanol and the recovery is examined by the peak intensity. The

Preparatory Investigations

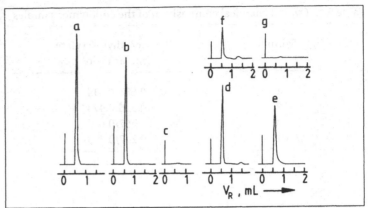

Fig. 5.17. Effect of ethanol content in the mobile phase on the elution of samples 2 *(a-c)* and 3 *(d-g)*; column temperature 50 °C; stationary phase:silica gel; mobile phase: chloroform-ethanol (% by volume); a 97:3, b 98:2, c 99:1, d 95:5, e 96:4, f 97:3, g 98:2

chromatograms for different eluent compositions are presented in Fig. 5.17.

Sample 2 does not elute from the column when the ethanol content in the mobile phase is 1% by volume. At 2% ethanol concentration, about 80% of the total sample are eluted and at 3% ethanol the copolymer is completely eluted from the column. Sample 3 is retained in the column with the mobile phase containing ethanol up to 2%, and the peak height increases as the ethanol content is increased from 3% to 5% by volume. Copolymer 3 is completely eluted in mobile phases containing 5% or more of ethanol.

The incomplete elution of the copolymers in Fig. 5.17 (b, e and f) is caused by difference in composition and not in molar mass; e.g. the fraction of copolymer S-EA 3 eluting with the mobile phase chloroform-ethanol 97:3% by volume (Fig. 5.17 f) had a higher styrene content than that retained on the column. This means that the copolymers may have some associated CCD.

Separations

Based on the fact that each copolymer is eluted only at a specific mobile phase composition, a mixture of copolymers of different chemical composition can be separated. Since a higher content of EA in the copolymer requires a higher amount of ethanol in the mobile phase, by stepwise increasing the concentration of ethanol a separation with respect to the EA content in the copolymer can be carried out. The separation of a mixture of copolymers 1–4 according to composition is shown in Fig. 5.18. In this case, a linear gradient is used for elution. As can be seen in the Figure, sepa-

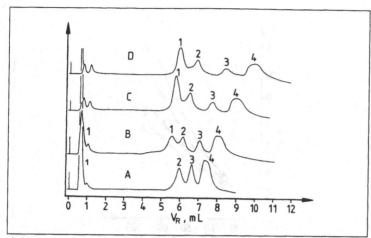

Fig. 5.18. LAC chromatograms of a mixture of copolymers 1–4; column temperature: 40 °C (**A**), 50 °C (**B**), 60 °C (**C**) and 70 °C (**D**); stationary phase:silica gel; mobile phase: chloroform-ethanol, linear gradient

ration is influenced significantly by temperature. When increasing the temperature, separation is improved and all fractions are properly retained.

The first peak in all chromatograms is the solvent peak. At a column temperature of 40 °C, fraction 1 is not retained and appears at the interstitial volume. When the temperature is increased to 50 °C, a part of fraction 1 is properly retained and a part is eluted at the interstitial volume, most probably due to slight differences in chemical composition. A proper retention of all fractions is achieved at column temperatures of 60 °C and 70 °C. In these cases, the desired separation with respect to the EA content takes place.

The quantitative analysis of the chromatograms yields the weight fractions of the sample components. The determination of the chemical composition of the fractions is based on the relation between EA content of the copolymer and retention volume. With the used calibration samples of known composition, a calibration curve EA content vs retention volume may be constructed and used for quantification (see Fig. 5.19 for S-EA copolymers).

Evaluation

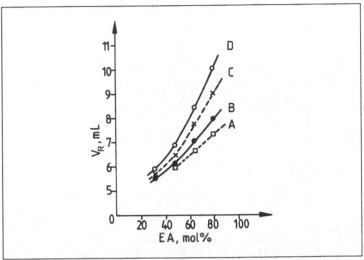

Fig. 5.19. Plots of retention volume vs EA content for S-EA copolymers at different column temperatures: 40 °C **(A)** 50 °C **(B)** 60 °C **(C)** and 70 °C **(D)**

References

1. GIDDINGS JC (1965) Dynamics of Chromatography. Marcel Dekker, New York
2. GLÖCKNER G (1991) Gradient HPLC of Copolymers and Chromatographic Cross-Fractionation. Springer, Berlin Heidelberg New York
3. GLÖCKNER G (1982) Polymercharakterisierung durch Flüssigchromatographie. Deutscher Verlag der Wissenschaften, Berlin
4. GLÖCKNER G (1986) Adv Polym Sci 79:159
5. EISENBEISS F, DUMONT E, HENKE H (1978) Angew Makromol Chem 71:67
6. SNYDER LR, STADALIUS MA (1986) High Performance Liquid Chromatography 4:195
7. SNYDER LR (1980) High Performance Liquid Chromatography 1:207
8. JANDERA P, CHURACEK J (1985) Gradient Elution in Column Liquid Chromatography, Theory and Practice. J Chromatogr Library 31
9. GLÖCKNER G, VAN DEN BERG JHM (1987) J Chromatogr 384:135
10. GLÖCKNER G, STICKLER M, WUNDERLICH W (1989) J Appl Polym Sci 37:3147
11. GLÖCKNER G, MÜLLER AHE (1989) J Appl Polym Sci 38:1761
12. SATO H, TAKEUCHI H, TANAKA Y (1986) Macromolecules 19:2613
13. TERAMACHI S, HASEGAWA A, SHIMA Y, AKATSUKA M, NAKAJIMA M (1979) Macromolecules 12:992
14. GLÖCKNER G, VAN DEN BERG JHM (1986) J Chromatogr 352:511
15. AUGENSTEIN M, STICKLER M (1990) Makromol Chem 191:415
16. GLÖCKNER G (1987) J Chromatogr 403:280
17. ABBOTT SR, TUSA J (1983) J Liquid Chromatogr 6:77

18. Scott RPW (1986) Liquid Chromatography Detectors. Elsevier, Amsterdam

19. Yeung ES (1986) Detectors for Liquid Chromatography. Wiley, New York

20. Shafer KH, Pentoney SL, Griffiths PR (1984) J High Res Chromatogr Chromatogr Commun 7:707

21. Willis JM, Dwyer JL Wheeler L (1993) 6th Int Symp Polym Anal Char Crete Greece Proc p 111

22. Hamielec AE, Ouano AC, Nebenzahl LL (1978) J Liquid Chromatogr 1:527

23. Haney MA (1985) J Appl Polym Sci 30:3037

24. Yau WW, Abboutt SD, Smith GA, Keating MY (1987) ACS Sym Ser 352:80

25. Glöckner G, Stickler M, Wunderlich W (1987) Fresenius Z Anal Chem 328:76

26. Okada T (1992) J Chromatogr 609:213

27. Trathnigg B, Thamer D, Yan X, Kinugasa S (1993) J Liquid Chromatogr 16:2439

28. Alexander JN, McNally ME, Rogers J (1985) J Chromatogr 318:289

29. Rissler K, Fuchslueger U, Grether HJ (1994) J Liquid Chromatogr 17:3109

30. Eisenbeiss F, Dumont E, Henke H (1978) Angew Makromol Chem 71:67

31. Hatada K, Ute K, Okamoto Y, Imanari M, Fuji N (1988) Polymer Bull 20:317

32. Albert K, Schlotterbeck G, Braumann U, Händel H, Spraul M, Krack G (1995) Angew Chem 107:1102

33. Stolyhwo A, Colin H, Guiochon G (1983) J Chromatogr 265:1

34. Mourey TH, Oppenheimer LE (1984) Anal Chem 56:2427

35. Oppenheimer LE, Mourey TH (1985) J Chromatogr 323:297

36. Augenstein M, Müller MA (1990) Makromol Chem 191:2151

37. Webster OW, Hertler WR, Sogah DY, Farnham WB, Rajan-babu TV (1983) J Am Chem Soc 105:5706

38. Hertler WR, Sogah DY, Webster OW, Cohen GM (1987) Macromolecules 20:1473

39. Webster OW, Hertler WR, Sogah DY, Farnham WB, Rajan-babu TV (1984) J Macromol Sci Chem A21:943

40. Glöckner G (1988) Chromatographia 25:854

41. Mori S, Mouri M (1989) Anal Chem 61:2171

42. Danielewicz M, Kubin M (1981) J Appl Polym Sci 26:951

43. Glöckner G, van den Berg JHM (1986) J Chromatogr 352:511

44. Glöckner G, van den Berg JHM, Meijerink NLJ, Scholte TG, Koningsveld R (1984) Macromolecules 17:962

45. Mori S, Uno Y, Suzuki M (1986) Anal Chem 58:303

46. Mori S, Uno Y (1987) Anal Chem 59:90

6 Liquid Chromatography at the Critical Point of Adsorption

6.1 Peculiarities

Liquid chromatography at the critical point of adsorption relates to a chromatographic situation, where the entropic and enthalpic interactions of the macromolecules and the packing compensate for each other. The Gibbs free energy of the macromolecule does not change when entering the pores of the stationary phase

$$\Delta H = T\Delta S \tag{6.1}$$

$$\Delta G = 0 \tag{6.2}$$

The distribution coefficient K_d is unity, regardless of the size of the macromolecules, and all macromolecules of equal chemical structure elute from the chromatographic column in one peak. For the description of this phenomenon the term "chromatographic invisibility" is used, meaning that the chromatographic behaviour is not directed by the size but by the inhomogeneities (chemical structure) of the macromolecules.

As the Gibbs free energy in general is influenced by the length of the polymer chain and its chemical structure, contributions ΔG_i for the polymer chain and ΔG_j for the heterogeneity may be introduced.

$$\Delta G = \Sigma n_i \Delta G_i + n_j \Delta G_j \tag{6.3}$$

For a homopolymer chain without chemical heterogeneity ΔG equals the contribution of the polymer chain.

$$\Delta G = \Sigma n_i \Delta G_i \tag{6.4}$$

At the critical point of adsorption of the polymer chain, however, its contribution ΔG_i becomes zero and chromatographic behaviour is exclusively directed by the heterogeneity.

$$\Delta G = \Sigma n_j \Delta G_j \tag{6.5}$$

Under such chromatographic conditions it is possible to determine the heterogeneities of the polymer chain selectively and without any influence of the polymer chain length.

The transition from one to another chromatographic separation mode can be accomplished by changing the temperature or the composition of the mobile phase. However, from the experimental point of view it is much more favourable to change the composition of the mobile phase rather than the temperature. In this case the balance of entropic and enthalpic interactions is achieved much more easily.

To explain the experimental treatment of liquid chromatography at the critical point of adsorption, the individual steps required to determine the critical point of poly(ethylene glycol) are discussed. The critical point is usually obtained by investigating a number of samples of different molar mass, preferably calibration standards, in eluents of varying composition. The chromatographic system must be selected such that the interaction of the end groups differs from the interactions of the polymer chain. In the case of poly(ethylene glycol), a reversed-phase stationary phase based on octadecyl-modified silica gel (RP-18) is selected.

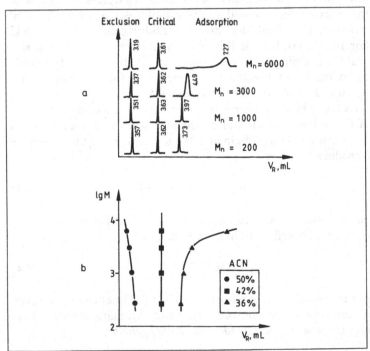

Fig. 6.1. Chromatograms **(a)** and critical diagram molar mass vs elution volume **(b)** of poly(ethylene glycols) at different mobile phase compositions, stationary phase: Nucleosil RP-18, mobile phase: acetonitrile-water (First published in LC-GC-International, Vol. 5, No. 2, 1992)

Figure 6.1 shows the chromatographic behaviour of a number of poly(ethylene glycols) on a RP-18 column using a mobile phase comprising mixtures of acetonitrile-water [1]. When the acetonitrile concentration in the mobile phase is >42% by volume, the retention time decreases as the molar mass of the sample increases. Accordingly, retention corresponds to a size exclusion mode. In mobile phases, comprising a higher amount of water, a different retention behaviour is obtained. At an acetonitrile concentration <42% by volume in the mobile phase, the retention time of the samples increases with increasing molar mass, indicating that the system is in the adsorption mode. At exactly 42% by volume of acetonitrile in the mobile phase, the critical point of adsorption of poly(ethylene glycol) is obtained. At this point all samples, regardless of their molar mass, elute at the same retention time and a straight line in parallel to the molar mass axis is obtained in the molar mass vs retention time plot.

6.2 Equipment and Materials

Chromatographic experiments in the critical mode of liquid chromatography are carried out on conventional liquid chromatographic equipment, comprising an LC pump, a sample introduction device, a chromatographic column, a detector, and a data processing unit (see Chapter 3). Depending on the specific separations to be conducted, e.g., separation with respect to functional end groups of telechelics or separation with respect to the individual blocks of a block copolymer, different stationary and mobile phases may be used. In principle, the stationary and mobile phases are not different from what is used in SEC and adsorption chromatography. However, they must be selected such that by varying the composition of the mobile phase, the entropic and enthalpic interactions between the macromolecules and the stationary phase may be influenced. In particular, it must be possible to operate simultaneously in adsorption, partition, and size exclusion modes on one column just by changing the composition of the mobile phase.

Stationary Phases. As the critical mode is a specific type of interaction chromatography, the packing must have active sites at the inner and outer surface, where interaction with the macromolecules takes place according to their polarity and hydrodynamic volume. The stationary phase must withstand high pressure, and the particle and pore size must not change when the mobile phase composition changes. Therefore, stationary phases of bare or modified silica are usually preferred, although crosslinked polymeric gels may be used as well. Criteria for the quality of stationary phases are:

- size and shape of the particles
- specific surface
- pore size and pore size distribution
- chemical stability
- pressure stability.

Mobile Phases. In the critical mode of liquid chromatography usually binary eluents are used. They comprise a thermodynamically good and a thermodynamically poor solvent. The thermodynamically good solvent controls the entropic interactions in the chromatographic system, whereas the poor solvent, by favouring adsorptive interactions, is responsible for changes in adsorption enthalpy. Thus, by changing the ratio of good to poor solvent in the mobile phase, the entropic and enthalpic interactions may be balanced against each other. Accordingly, the main requirements for the mobile phase are:

- solubility of the sample
- controlled interactions with the solute and the packing
- suitability for a specific detector
- chemical stability
- miscibility of the components
- viscosity
- environmental safety.

Detectors. As for all detectors in liquid chromatography, sensitivity is a key parameter. In the critical mode of liquid chromatography any detector may be used, in particular, universal detectors such as the refractometer and the evaporative light scattering detector. For selective detection, UV photometers are commonly used, but IR, viscosity and light scattering detectors may be used as well.

One specific feature of liquid chromatography at the critical point of adsorption is the use of binary mobile phases. Depending on the composition of the mobile phase, the physical properties such as UV response, refractive index and viscosity may change. This must be taken into account when selecting a certain detector. In addition, in the vicinity of the critical point preferential adsorption of one component of the mobile phase may occur. Due to the different solvating power of the mobile phase components, one of them may be preferentially adsorbed to the macromolecules in solution. This causes an increase of the volume frac-

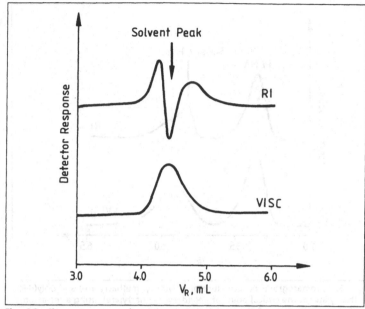

Fig. 6.2. Chromatograms of poly(methyl methacrylate) at the critical point of adsorption, using a refractometer (RI) and a viscometer (VISC) as detector

tion of the elution peak and a local depletion of this particular component in the mobile phase. This may give rise to a change of the area of the elution peak and the appearance of a more or less pronounced solvent peak. As both the solvent and the polymer peak at the critical point of adsorption appear close to $K_d = 1$, an overlapping of both peaks in some cases cannot be avoided.

Figure 6.2 shows the chromatographic behaviour of a poly(methyl methacrylate) calibration standard in the critical mode of liquid chromatography. In the SEC mode the polymer peak and a small solvent peak are well separated from each other, whereas in the critical mode both peaks completely overlap. The solvent peak has a negative response and for the polymer peak neither the peak maximum nor the peak area can be determined. If a viscometer instead of the refractometer is used for the detection, the mobile phase response is negligible compared to the polymer response. Therefore, the polymer peak can be detected without interference by solvent peaks (see Fig. 6.3). To avoid interferences by solvent peaks, evaporative light scattering detectors and IR detectors with a solvent-evaporation interface can be used as well.

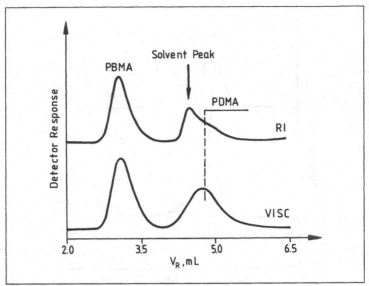

Fig. 6.3. Chromatograms of a mixture of poly(butyl methacrylate) and poly(decyl methacrylate) at the critical point of poly(decyl methacrylate), using a refractometer (RI) and a viscometer (VISC) as detector

6.3 Data Acquisition and Processing

For data acquisition and processing conventional equipment is used similar to SEC and adsorption chromatography. Also, calibration and quantification can be conducted similar to the corresponding procedures in SEC and adsorption chromatography. Provided that one of the components of the sample elutes in the size exclusion mode, the corresponding SEC calibration curve may be used for determining the molar mass distribution of this component. However, in all cases one must take care of an appropriate detection system, assuring that the solvent peak does not interfere the quantification. For functionality type separations usually well resolved peaks for individual functionality fractions are obtained. The response factors of the different functionality fractions must be determined for each detector system. Using these response factors, quantification may be carried out via determining the peak area or peak height of a particular elution peak.

6.4 Separation of Functional Homopolymers

6.4.1 Principles and Limitations

The use of liquid chromatography for the analysis of functional homopolymers was proposed by Muenker and Hudson [2] and Evreinov [3]. With mixtures of solvents of increasing solvent strength, hydrogenated oligobutadienes with hydroxyl and carboxyl terminal groups were separated according to their functionality types. A practically complete separation of non-, mono- and bifunctional fractions could be achieved [2]. For poly(diethylene glycol adipate) it was shown that on silica gel a separation according to the number of terminal hydroxyl groups takes place [3].

With respect to the chromatographic system, three different cases can be encountered(see Fig. 6.4):

1. The functional groups do not manifest themselves chromatographically. The calibration curves for fractions of different functionality coincide, and the molar mass distribution obtained is a superposition of the distributions of the functionality fractions. This situation is encountered in ideal SEC and does not allow the determination of the functionality type distribution (FTD).

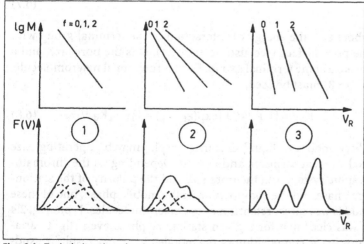

Fig. 6.4. Typical situations in chromatography of functional homopolymers; for explanations of 1, 2, and 3, see text (Reprinted from Ref. [4] with permission of Springer-Verlag)

2. The functional groups manifest themselves only slightly. The calibration curves do not coincide, but because of the molar mass distributions the elution zones overlap. This may be the case in real SEC and gives a first indication of the presence of a FTD. The poor separation of the functionality fractions, however, does not allow to determine FTD quantitatively.

3. Complete separation of the functionality fractions can be obtained, when the interaction of the terminal groups with the adsorbent is much stronger than that of the polymer chain. This situation is encountered in ideal liquid chromatography at the critical point of adsorption and enables the quantitative determination of FTD.

The simplest type of a functional homopolymer is a telechelic, i.e., a macromolecule containing terminal groups. The presence of these groups causes a difference in the interaction of the macromolecule with the packing, in particular with the pore surface. If a binary mobile phase is chosen, corresponding to the critical point of adsorption for the polymer chain, the polymer chain itself will not contribute to any variations of ΔG. Accordingly, for a non-functional macromolecule the distribution coefficient will be unity.

$$K_d^{(0)} = 1 \qquad (6.6)$$

For a monofunctional macromolecule, the contribution to ΔG will only be made by the terminal group, and the distribution coefficient may be calculated from equation (6.7) [5].

$$K_d^{(1)} = 1 - 2a/D + 2a/D \exp(\varepsilon_f - \varepsilon_c) = 1 + 2a/D \left[\exp(\varepsilon_f - \varepsilon_c) - 1 \right] \qquad (6.7)$$

where ε_f is the energy of interaction of the terminal group with the pore surface (see also Section 2.4), D is the pore size, and a the size of the terminal group. For a bifunctional macromolecule, Eq. (6.8) must be used,

$$K_d^{(2)} = \{ 1 + 2a/D \left[\exp(\varepsilon_f - \varepsilon_c) - 1 \right] \}^2 = [K_d^{(1)}]^2 \qquad (6.8)$$

Three modes of liquid chromatography may be operating: size exclusion, adsorption, and critical. Depending on the chromatographic system, i.e., the pore size and the polarity of the stationary phase and the composition of the mobile phase, one of these modes will direct retention of the macromolecules. From Fig. 2.4 it was clear that for a given stationary phase even slight variations in the composition of the mobile phase may result in a sudden transition from one separation mode to another. The transi-

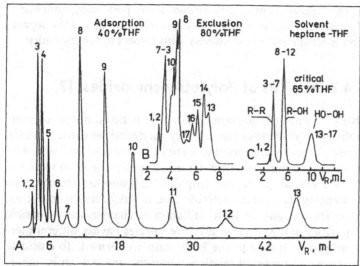

Fig. 6.5. Separation of hydroxy-terminated poly(butylene glycol terephthalate) as a function of the mobile phase composition; stationary phase: silica gel; mobile phase: tetrahydrofuran-heptane; peaks 1, 2 – solvent, peaks 3–7 non-functional homologues (n=0–4), peaks 8–12 monofunctional homologues (n=0–4), peaks 13–17 bifunctional homologues (n=0–4) (Reprinted from Ref. [4] with permission of Springer-Verlag)

tion through different separation modes by changing the composition of the mobile phase is illustrated for poly(butylene glycol terephtalate) in Fig. 6.5 [6].

At a mobile phase of THF-heptane 80:20% by volume, separation is accomplished according to the size exclusion mode. Within the same functionality, the large molecules elute first. At THF-heptane 40:60% by volume, retention volume increases with increasing molar mass and, therefore, the adsorption mode is operating. In both separation modes the effects of functionality and molar mass overlap. The most interesting feature is the separation at the critical point of adsorption. In this case, using a mobile phase of THF-heptane 65:35% by volume, separation of the sample into three fractions is obtained, which may be assigned to the functionality fractions:

– Peaks 3–7 $\quad CH_3O-OC-\left[\bigcirc-CO-O(CH_2)_4O\right]_n OC-\bigcirc-COOCH_3$

– Peaks 8–12 $\quad CH_3O-OC-\left[\bigcirc-CO-O(CH_2)_4O\right]_n H$

– Peaks 13–17 $\quad HO(CH_2)_4O-OC-\left[\bigcirc-CO-O(CH_2)_4O\right]_n H$

At the critical point of adsorption, a complete independence of V_R of the size of the macromolecules is obtained, and the separation is accomplished exclusively with respect to functionality.

6.4.2 Analysis of Poly(ethylene oxides) [7]

Aim

One of the most important classes of functional homopolymers is the class of alkyloxy and aryloxy poly(ethylene oxides) (PEO). These oligomers and polymers are in widespread use as surfactants. Depending on their molar mass and the chemical structure of the terminal groups the amphiphilic properties change, thus influencing the surface activity. Due to the different initiation, chain transfer and chain termination mechanisms and possible impurities in the reaction mixture, species having different terminal groups bound to the PEO chain are formed. To elucidate the structure-property relationship of the products, it is important to know the chemical structure and the number of the terminal groups in addition to the molar mass distribution.

Materials

Calibration Standards. Narrow-disperse poly(ethylene glycols) in the molar mass range of 200–10 000 g/mol.

Polymers. Technical alkoxy and aryloxy PEOs of the following average structure (see Table 6.1).

Equipment

Chromatographic System. Modular HPLC system comprising a Waters model 510 pump, a Rheodyne six-port injection valve and a Waters column oven.

Columns. Nucleosil RP-18 and RP-8 of Macherey-Nagel, all of 5 µm average particle size and 100 Å average pore diameter. Column size was 125x4 mm I.D. or 60x4 mm I.D.

Mobile Phase. Mixtures of acetonitrile and water, all solvents are HPLC grade.

Table 6.1. Average structure of the PEOs, as supplied by the manufacturer

Number	Sample	Average Structure
1	C_{10}-PEO	$C_{10}H_{21}(OCH_2CH_2)_7OH$
2	C_{12}-PEO	$C_{12}H_{25}(OCH_2CH_2)_7OH$
3	C_{13}-PEO	$C_{13}H_{27}(OCH_2CH_2)_8OH$
4	C_{13},C_{15}-PEO	$C_{13}H_{27},C_{15}H_{31}(OCH_2CH_2)_7OH$
5	octylphenol-PEO	$C_8H_{17}C_6H_4(OCH_2CH_2)_6OH$
6	nonylphenol-PEO	$C_9H_{19}C_6H_4(OCH_2CH_2)_{10}OH$

Detectors. Waters differential refractometer R 401 or R 410 and Knauer UV/vis filter photometer or Waters model 486 tunable UV detector.

Column Temperature. 25 °C.

Sample Concentration. 0.5–4 mg/mL. All samples are dissolved in the mobile phase

Injection Volume. 20–50 µL.

As was described in Sections 2.4 and 6.1, separations according to the terminal groups must be carried out at chromatographic conditions, corresponding to the critical point of the polymer chain, in the present case PEO. The critical point of PEO is determined by running a set of poly(ethylene glycols) (PEG) of different molar masses at different mobile phase compositions. It was shown by Gorshkov et al. [5] that PEG, vinyl- and butyl-terminated PEO may be analysed, using a reversed-phase column RP-18 of 250 mm length, and acetonitrile-water as the mobile phase. For PEOs with longer hydrophobic chain ends, this column is not suitable due to very strong hydrophobic interactions. To avoid irreversible adsorption on the stationary phase, packing material of lower hydrophobicity, such as RP-8 or RP-4, or very short columns must be used. According to the classification of RP columns, RP-18, RP-8 and RP-4 indicate octadecyl-, octyl- and butyl-modified silica gel, respectively. *[marginal note: Preparatory Investigations]*

The critical diagram M vs retention time for a short RP-18 column of 60 mm length is shown in Fig. 6.6A. At acetonitrile concentrations greater than 47% by volume in the mobile phase retention corresponds to a size-exclusion mode whereas the adsorption mode is operating at acetonitrile concentrations less than 45% by volume. The critical point of adsorption is obtained at a mobile phase composition of acetonitrile-water 46:54% by volume, where regardless of the molar mass all calibration standards elute at the same retention time. In a similar procedure, the critical point for PEG may be determined on a reversed-phase RP-8, where the critical point is obtained at a mobile phase composition of acetonitrile-water 44:56% by volume (see Fig. 6.6B).

Separations of the alkoxy and aryloxy PEOs according to their functional end groups are carried out at the critical point of PEG. The stationary phase is Nucleosil RP-18, 60x4 mm I.D., column temperature is 25 °C, and mobile phase composition is acetonitrile-water 46:54% by volume. *[marginal note: Separations]*

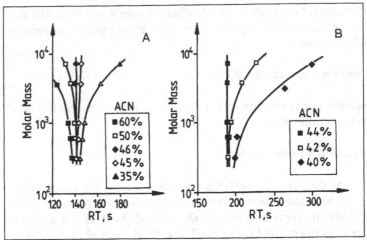

Fig. 6.6. Critical diagrams molar mass vs retention time of polyethylene glycol; stationary phase: Nucleosil RP-18 **(A)** and RP-8 **(B)**, 60x4 mm I.D.; mobile phase: acetonitrile-water

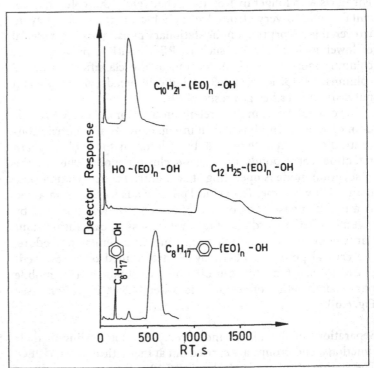

Fig. 6.7. Chromatograms of functional PEOs (samples 1, 2, 5) at the critical point of adsorption of PEG; stationary phase: Nucleosil RP-18, 60x4 mm I.D.; mobile phase: acetonitrile- water 46:54% by volume

The chromatograms of functional PEOs with different end groups are shown in Fig. 6.7. For the C_{10}- and C_{12}-PEOs (samples 1 and 2) two distinctively different fractions are obtained, resulting in a sharp peak at a retention time of about 35 s and a broad peak at higher retention times. For the octylphenol-PEO (sample 5), where the UV detector at a wavelength of 280 nm is used, this first peak is not obtained, indicating that the corresponding fraction has no UV activity. By comparison with a standard sample, the first peak is identified as PEG, which is known to be formed as an unwanted by-product.

In the case of the octylphenol-PEO, in addition to the main peak for the α-octylphenoxy-ω–hydroxy fraction two small peaks at retention times of 154 s and 303 s are detected. By comparison with the pure compound, the peak at 154 s is found to be octylphenol. The second peak at 303 s is an aryloxy PEO. From the smaller retention time compared to the main fraction, the second peak is assumed to be an alkylphenol-PEO with a smaller alkyl substituent, which might have been formed due to impurities in the starting octylphenol.

The quantitative determination of the PEG fraction (first peak in Fig. 6.7) is carried out using a concentration calibration curve of PEG vs refractive index response. The PEG content for the samples under investigation is in the magnitude of 1–3% by weight, which is typically for these types of commercial products.

It is known that reversed-phase columns separate molecules with respect to hydrophobicity. Accordingly, the retention time of the samples increases with increasing hydrophobicity of the terminal group. Therefore, the elution order with respect to the terminal group is

$$OH\ (PEG) << C_{10}H_{21} < C_{12}H_{25} < C_{13}H_{27} < C_{15}H_{31}$$

The change of retention times with composition of the mobile phase is given in Fig. 6.8. The dramatic increase of the retention times near the critical point suggests that PEOs with alkoxy end groups greater than C_{13} may not be eluted from the column in reasonable retention times. Indeed, $C_{15}H_{31}$-terminated PEO and samples with longer alkoxy end groups do not elute from the column within 60 min.

In order to decrease the hydrophobicity of the stationary phase and the retention times accordingly, an octyl-modified (RP-8) instead of an octadecyl-modified (RP-18) silica gel may be used. Owing to the lower hydrophobicity of the RP-8 packing material, the samples elute much faster from the column, compare Figs. 6.7 and 6.9.

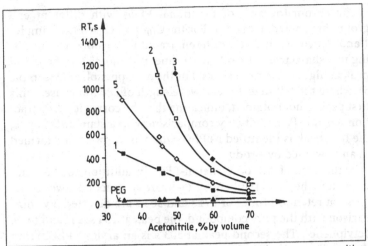

Fig. 6.8. Diagram of retention time of the main functional fraction vs composition of the mobile phase; for stationary and mobile phases see Fig. 6.7; numbers indicate sample numbers (see Table 6.1)

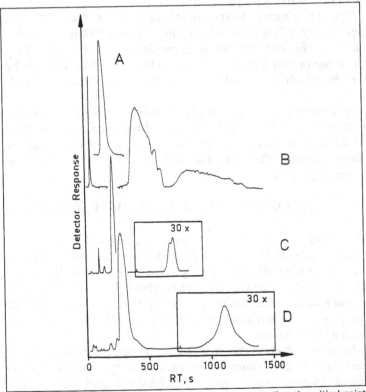

Fig. 6.9. Chromatograms of functional PEOs (samples 1, 4, 5, 6) at the critical point of adsorption of PEG; stationary phase: Nucleosil RP-8, 60x4 mm I.D.; mobile phase: acetonitrile- water 44:56% by volume; samples: C_{10}-PEO (A), C_{13},C_{15}-PEO (B), octylphenol-PEO (C), nonylphenol-PEO (D), RI detection (A, B), UV detection (C, D)

The critical point of PEG on the RP-8 column corresponds to a mobile phase composition of acetonitrile-water 44:56% by volume (see Fig. 6.6B). On this column all samples elute in a reasonable time. In addition to the previously detected peaks, for the aryloxy samples (samples 5 and 6) a third functionality fraction elutes, which obviously corresponds to the α,ω-diaryloxy species. The concentration of this third functionality fraction, however, is very low and can be detected in the reaction mixture only with a 30-fold sensitivity increase of the UV detector.

An important feature of the present chromatographic system is the possibility to modify separation by changing the stationary phase and/or the composition of the mobile phase. In some cases the chromatographic system can be modified in such a way that separation according to functionality *and* oligomer distribution becomes possible.

Modified Procedure

Figure 6.10 shows the separation of the samples on a 125 mm long RP-18 column using a mobile phase of acetonitrile-water

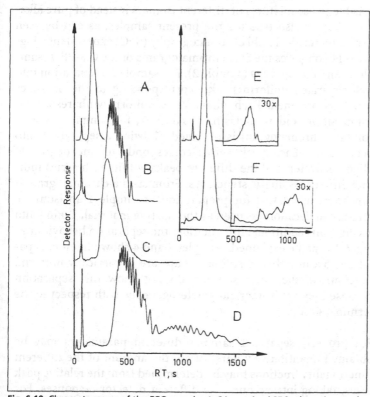

Fig. 6.10. Chromatograms of the PEO samples 1–6 in a mixed SEC-adsorption mode; stationary phase: Nucleosil RP-18, 125x4 mm I.D.; mobile phase: acetonitrile-water 70:30% by volume; samples 1 **(A)**, 2 **(B)**, 3 **(C)**, 4 **(D)**, 5 **(E)**, 6 **(F)**

70:30% by volume. In this case separation corresponds to a mixed SEC-adsorption mechanism. Inspection of the chromatograms reveals a separation into oligomer series for the C_{12}- and the C_{13},C_{15}-PEOs, whereas the other samples are not separated into oligomers. Regardless of the type of the functional group, in all cases a sufficient separation is obtained from the PEG and the diaryloxy fractions, thus allowing the determination of all functionality fractions. The proposed mixed SEC-adsorption mode is still very sensitive towards the chemical structure of the terminal groups. Even an increase of the substituent chain length by one methylene unit in the aryloxy PEOs (from octyl to nonyl in samples 5 and 6, respectively) results in a significant increase in retention time. With a difference of two methylene units in the terminal group a complete separation of the functional fractions is obtained (see C_{13}- and C_{15}-fractions of sample 4).

From PEO chemistry, it is known that fatty alcohols of different structure are used as starting materials. They may not only differ in the length of the alkyl chain but also in its isomeric form (n-alkanol vs iso-alkanol). From economic reasons it is often more feasible to use mixtures of different isomers instead of pure alkanols. This is also true for the present samples, as can be seen from supercritical fluid chromatography (SFC) experiments. Figure 6.11 compares the SFC chromatograms of the C_{10}-PEO (sample 1) and the C_{12}-PEO (sample 2). For sample 2 a separation into well-separated uniform peaks, corresponding to the oligomer series, is obtained. Each oligomer is uniform with respect to composition and isomerism. In contrast, for sample 1 a very complex chromatogram is obtained, showing a separation into peak series. Each peak series corresponds to one degree of oligomerization and the different peaks within each series indicate different isomeric structures. From such a chromatogram it can be assumed that for preparation of sample 1 an isomeric mixture of decanol was used as the starting material. Taking into account this additional information, the separation behaviour in Fig. 6.10 can be explained. Samples 2 and 4 show oligomer separations because their terminal groups are isomerically uniform. In contrast, the other samples do not show this separation because they are isomerically heterogeneous with respect to the terminal group.

Evaluation

For properly separated samples different parameters may be obtained quantitatively. First of all, the amount of the different functionality fractions may be determined from the relative peak areas, taking into account the different detector responses. For the determination of the molar mass distribution of each functionality fraction, preparative separations may be carried out

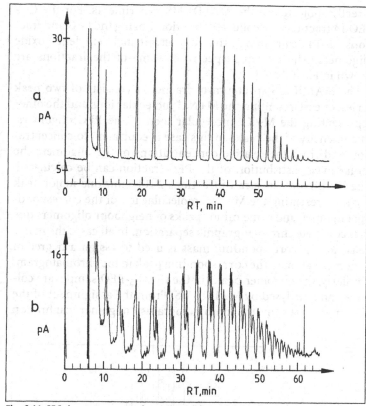

Fig. 6.11. SFC chromatograms of samples 1 **(b)** and 2 **(a)** indicating the different isomeric structures of the terminal groups

and the resulting fractions may be investigated by SEC or any other method for molar mass determination.

For samples that can be separated with respect to functionality and oligomer distribution (see Fig. 6.10, samples 2 and 4), a different approach may be used. From the analytical separation, fractions of the single oligomers are collected and subjected to mass spectrometry for identification. In particular, matrix-assisted laser desorption/ionization mass spectrometry (MAL-DI-MS) is a useful new technique for the analysis of polymer samples with respect to chemical structure and molar mass [9–11].

In order to assign the oligomer peaks in the chromatograms of samples 2 and 4, a chromatographic separation is conducted and the oligomer fractions are collected, resulting in amounts of 5–20 ng substance per fraction in the mobile phase; the solutions are directly subjected to the MALDI-MS experiments. For the C_{12}-PEO 14 fractions are collected, fraction 1 being the PEG and fractions 2–14 containing the C_{12}-terminated ethylene oxide oligomers. The resulting spectra of some of the fractions are shown in Fig. 6.12 [7].

The MALDI-MS spectrum of fraction 1 consists of two peak series, one representing the $M+Na^+$ molecular ions and the other representing the $M+K^+$ molecular ions of the PEG oligomers. The intensity of the peaks in this case is equivalent to concentration and from the relative concentrations of the oligomers the molar mass distribution of the PEG fraction can be calculated. The MALDI-MS spectra of fractions 2–14 show one major peak each, representing the $M+Na^+$ molecular ion of the corresponding oligomer, and some minor peaks of neighbour oligomers due to incomplete chromatographic separation. In all cases the major peak and its corresponding mass is used to assign a degree of oligomerization to the corresponding peak in the chromatogram. Similarly, the oligomer peaks in the C_{13}, C_{15}-PEO sample are collected and analysed by MALDI-MS. From the assignment of the oligomer peaks in the chromatograms, oligomer calibration

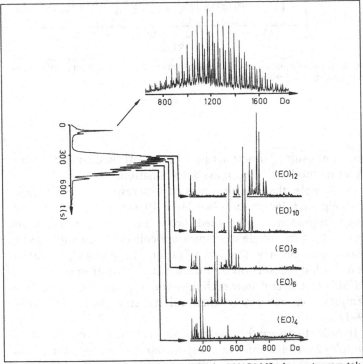

Fig. 6.12. Identification of fractions of C_{12}-PEO by MALDI-MS, chromatogram taken from Fig. 6.10

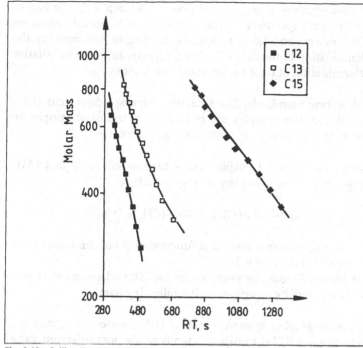

Fig. 6.13. Calibration curves molar mass vs retention time for the functionality fractions of samples 2 and 4; for corresponding chromatograms see Fig. 6.10

curves molar mass vs retention time may be obtained for the C_{12}-, C_{13}- and C_{15}-fractions (see Fig. 6.13). With these calibration curves, the average molar masses and the polydispersities of the functionality fractions are calculated.

6.4.3 Analysis of Aliphatic Polyesters [12]

Polyesters from adipic acid and 1,6-hexanediol are manufactured for a wide field of applications with an output of thousands of tons a year. They are intermediates for the manufacture of polyurethanes, and their functionality type distribution is a major parameter in affecting the quality of the final products. In particular, non-reactive cyclic species are responsible for the "fogging effect" in polyurethane foams.

The determination of the FTD of polyesters was proposed by a number of authors. Vakhtina [13] separated aliphatic polyesters by thin layer chromatography, and Filatova et al. [14, 15] investigated polyesters by gradient elution liquid chromatography and

Aim

chromatography at the critical point of adsorption. The aim of the present experiment is the separation of technical polyesters of adipic acid and 1,6-hexanediol according to functionality, the identification of the functionality fractions, and the quantitative determination of their molar mass distributions.

Materials

Calibration Standards. Since narrow-disperse adipic acid-hexanediol polyester samples are not available, technical samples in the molar mass range of 1000–6000 g/mol are used.

Polymers. Technical adipic acid – hexanediol polyesters (AH-polyesters) of the following average structure

$$HO-[-OC-(CH_2)_4-COO-(CH_2)_6-O-]_n-H$$

The average molar masses and functionalities of the samples are summarized in Table 6.2.

Additional model polyesters with specific end groups were prepared and will be discussed in the following part.

Equipment

Chromatographic System. Modular HPLC system comprising a conventional HPLC pump, a Rheodyne six-port injection valve and any conventional HPLC column oven.

Columns. Critical chromatography: Silica gel of Tessek (Czechoslovakia), 7 µm average particle size and 120 Å average pore diameter. Column size was 250x4 mm I.D. SEC: Two-column set of PL-gel (GB), 5 µm average particle size and 50 Å and 100 Å average pore diameter. Column size was 300x8 mm I.D.

Table 6.2. Molecular parameters of AH-polyesters as supplied by the manufacturer

Sample	M_n	OH-groups/molecule
PE 1	930	1.83
PE 2	1.300	1.78
PE 3	1.650	1.69
PE 4	2.400	1.61
PE 5	1.020	1.98
PE 6	1.400	1.86
PE 7	1.810	1.86
PE 8	2.220	1.73
PE 9	5.940	1.64

Mobile Phase. Critical Chromatography: Mixtures of acetone and hexane, SEC: acetone, all solvents are HPLC grade.

Detectors. ERC 7511 differential refractometer.

Column Temperature. 30 °C.

Sample Concentration. 0.5–5 mg/mL. All samples are dissolved in the mobile phase.

Injection Volume. 20–100 µL.

The critical point of adsorption of the polymer chain must be determined. As AH-polyesters may have different terminal groups such as HO-OH, HO-COOH and HOOC-COOH, they can differ in terms of a wide range of polarity. Therefore, silica gel and reversed phases may be tested as stationary phases. However, reversed phases were found to be unsuitable due to insufficient selectivity. In addition, the mobile phase composition corresponding to the critical point was found to be close to the precipitation point, causing partial precipitation of the higher molar mass samples on the column. With silica gel and a mobile phase of acetone-hexane, good solubility of the samples is obtained. Testing silica gels of different pore sizes, optimum resolution and peak shape is obtained with an average pore diameter of about 100 Å.

Preparatory Investigations

The critical diagram molar mass vs retention time for the silica gel 120 Å is shown in Fig. 6.14. The critical point of adsorption is obtained at a mobile phase composition of acetone-hexane 51:49% by volume, where regardless of the molar mass, all calibration samples elute at the same elution volume. At higher concentrations of acetone in the mobile phase the SEC mode is operating, whereas at lower concentrations of acetone in the mobile phase separation corresponds to the adsorption mode.

Separations of the AH-polyesters given in Table 6.2. according to their terminal groups are carried out at the critical point of adsorption, as indicated in Fig. 6.14. The packing is silica gel, 250x4 mm I.D., column temperature is 30 °C, and the mobile phase composition is acetone-hexane 51:49% by volume.

Separations

The critical chromatograms of AH-polyesters with different molar masses are given in Fig. 6.15. In all chromatograms one major peak at a retention time of about 14 min is obtained. In addition, a number of elution peaks of lower intensity indicate other functionality fractions. In total, up to nine elution peaks may be identified in the chromatograms. It is known from the

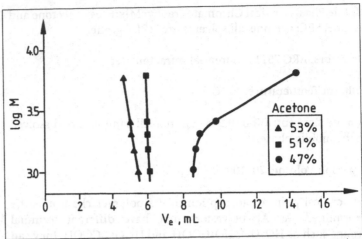

Fig. 6.14. Critical diagram molar mass vs elution volume of AH- polyesters; stationary phase: Tessek silica gel, 250x4 mm I.D.; mobile phase: acetone-hexane (Reprinted from Ref. [12] with permission of Marcel Dekker. Inc., USA)

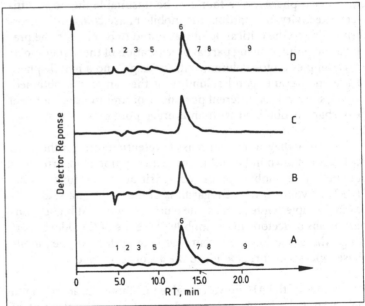

Fig. 6.15. Chromatograms of AH-polyesters of different molar masses; A-PE1, B-PE2, C-PE3, D-PE4; stationary phase: Tessek silica gel, 250x4 mm I.D.; mobile phase: acetone-hexane 51:49% by volume (Reprinted from Ref. [12] with permission of Marcel Dekker. Inc., USA)

hydroxy and acid numbers that the samples mainly are hydroxy-terminated AH-polyesters. Accordingly, the major peak (peak 6) can be assigned to the α,ω-dihydroxy species

$$HO\text{-}(CH_2)_6\text{-}O\text{-}[\text{-}OC\text{-}(CH_2)_4\text{-}COO\text{-}(CH_2)_6\text{-}O\text{-}]_n\text{-}H$$

Table 6.3. Peak assignment of Fig. 6.15

$$R^1\text{-}[\text{-OC-}(CH_2)_4\text{-COO-}(CH_2)_6\text{-O-}]_n\text{-}R^2$$

Fraction	Symbol	R^1	R^2
1	Alk-Alk	$CH_2=CH(CH_2)_4$	X
2	cycles	–	–
3	Alk-OH	$CH_2=CH(CH_2)_4$	-H
4	HOOC-COOH	HO-	$\text{-OC-}(CH_2)_4\text{-COOH}$
5	HOOC-OH	HO-	-H
6	HO-OH	$HO\text{-}(CH_2)_6\text{-O-}$	-H
9	1,6-hexanediol		

X: $\text{-OC-}(CH_2)_4\text{-COO-}(CH_2)_4\text{-CH=CH}_2$

For the assignment of the other peaks a number of model polyesters may be used, which have the same AH-polyester chain and specific end groups. By comparison of the chromatograms of these functionally uniform AH-polyesters with the peaks in Fig. 6.15, the assignment can be made as shown in Table 6.3[12].

Peaks 7 and 8 are obtained due to the formation of ether structures in the polyester samples.

$$HO\text{-}R^1\text{-OH} + HO\text{-}R^2\text{-OH} \rightarrow HO\text{-}R^1\text{-O-}R^2\text{-OH}$$

where R^1 and R^2 represent AH-polyester chains.

The amount of the different functionality fractions may be determined from the relative peak areas, taking into account the different detector responses. These data can be correlated with the hydroxy and the acid numbers of the total samples.

The molar mass distributions of the funtionality fractions may be determined by preparatively separating the fractions and subjecting them to SEC. The SEC chromatograms of fractions 1–9 are summarized in Fig. 6.16. For a number of fractions oligomer separations are obtained, which can be used to calibrate the SEC system (compare similar approach in Section 6.3.2). The SEC calibration curves for the functionality fractions 1, 2, 4–6 are given in Fig. 6.17. For functionality fractions 1 and 2 virtually the same calibration curve is obtained. The calibration curves for fractions 4–6 are very similar, but differ remarkably from the calibration

Evaluation

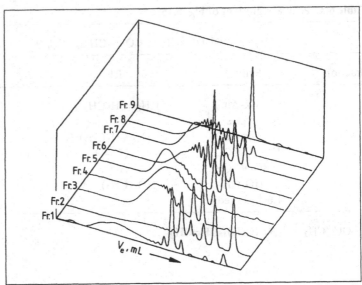

Fig. 6.16. SEC chromatograms of fractions 1–9 taken from separation of AH-poly-ester; stationary phase: PL-gel, 300x8 mm I.D.; mobile phase: acetone (Reprinted from Ref. [12] with permission of Marcel Dekker. Inc., USA)

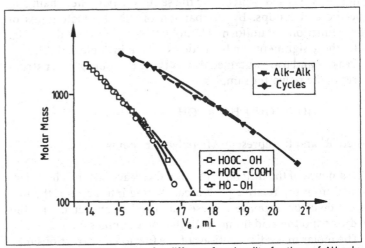

Fig. 6.17. SEC calibration curves for different functionality fractions of AH-poly-ester

curve of fractions 1 and 2. This clearly indicates that differences in end-group functionality have a strong effect on the SEC behaviour and must be considered when investigating this type of samples by SEC.

6.4.4 Analysis of Benzyloxy-terminated Poly(1,3,6-trioxocane)s

Polymers with properties adapted particularly to a specific applica-   Aim
tion are increasingly produced via macromonomers and telechelics
[16, 17]. Macromonomers are oligomers with a functional group at
one chain end, telechelics have α,ω-functional groups. The final
polymers arise from the reaction of these functional groups.

Macromonomers and telechelics can be synthesized by cationic ring-opening polymerization of oxygen-containing heterocycles. In the presence of hydroxy-functional substances, the formation of hydroxy-terminated oligomers takes place in high
yield. The synthesis of benzyloxy-terminated poly(1,3,6-trioxo-
cane)s can be carried out by reacting 1,3,6-trioxocane in the
presence of benzyl alcohol.

1

Owing to a number of secondary reactions, in addition to α-ben-
zyloxy-ω-hydroxy oligomers *1* the functionality fractions *2-4* are
obtained.

2

3

4

The analysis of the resulting complex reaction mixture is complicated, and SEC experiments alone are not sufficient for a complete characterization of the system. A possible way is the chromatographic separation according to the functionality type and a
subsequent analysis of the different functionality fractions.

Calibration Standards. Since narrow disperse poly(1,3,6-trioxo-   Materials
cane) (PTO) is not available commercially, a number of α,ω-
dihydroxy PTOs in the molar mass range of 700–71 000 g/mol
were prepared by cationic polymerization [18].

Polymers. Benzyloxy-terminated PTOs were prepared in a similar way, adding benzyl alcohol to the reaction mixture [18]. Sam-

ple A had a hydroxyl number of 164 mg KOH/g and a number-average molar mass of 385 g/mol. Sample B had a hydroxyl number of 137 mg KOH/g and a number average molar mass of 520 g/mol.

Equipment

Chromatographic System. Modular HPLC system comprising a conventional HPLC pump, a Rheodyne six-port injection valve and any conventional HPLC column oven.

Columns. Critical chromatography: C_{18}-modified silica gel, 10 mm average particle size and 300 Å average pore diameter. Column size was 250x4.6 mm I.D.. SEC: Two column-set of styrene-divinylbenzene gel with exclusion limits of 12 000 and 4000 g/mol, 10 mm average particle size. Column size was 300x8 mm I.D.

Mobile Phase. Critical chromatography: Mixtures of acetonitrile and water, SEC: tetrahydrofuran, all solvents are HPLC grade.

Detectors. UV/VIS detector SPD-6AV of Shimadzu and Waters 410 differential refractometer.

Column Temperature. 25 °C

Sample Concentration. 3 mg/mL for critical chromatography and 0.5 mg/ml for SEC. All samples are dissolved in the mobile phase.

Injection Volume. 20 μL

Preparatory Investigations

Similar to the previous investigations, in a first set of experiments the critical point of adsorption of the polymer chain must be determined. This is done by running a number of PTOs of different molar masses on a chromatographic system comprising a RP-18 reversed stationary phase and acetonitrile-water as the mobile phase. The critical diagram molar mass vs retention time is given in Fig. 6.18.

The critical point of adsorption for PTO is obtained at a mobile phase composition of acetonitrile-water 49.5:50.5% by volume. At higher concentrations of acetonitrile in the mobile phase, the SEC mode is operating, and at lower acetonitrile concentration separation corresponds to the adsorption mode.

Separations

The chromatograms of samples A and B representing a separation according to functionality are shown in Fig. 6.19. In both chromatograms four elution peaks may be distinguished; see

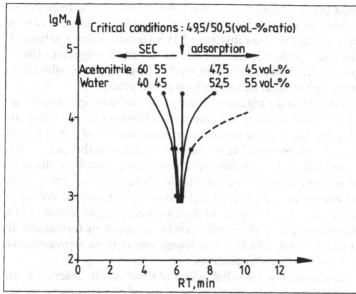

Fig. 6.18. Critical diagram molar mass vs retention time of α,ω-dihydroxy PTOs; stationary phase: RP-18, 250x4.6 mm I.D.; mobile phase: acetonitrile-water (Reprinted from Ref. [18] with permission of Hüthig & Wepf Publishers, Switzerland)

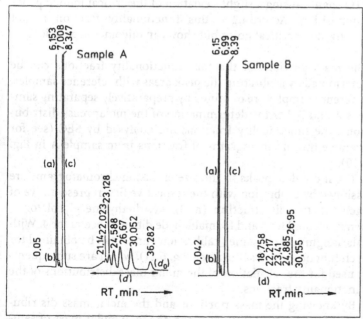

Fig. 6.19. Functionality chromatograms of samples A and B, stationary phase: RP-18, mobile phase: acetonitrile-water 49.5:50.5% by volume (Reprinted from Ref. [18] with permission of Hüthig & Wepf Publishers, Switzerland)

peaks (a)-(d). Since the elution time of peak (a) corresponds to the elution time of the calibration standards, this peak can be assigned to the α,ω-dihydroxy PTOs (functionality fraction 2). The small peak (b) corresponds to the cyclic oligomers (functionality fraction 3) as can be verified by comparison with chromatographic behaviour of various crown ethers.

Peak (c) has the highest intensity and can be assigned to the α-benzyloxy-ω–hydroxy oligomers (functionality fraction 1). Benzyl alcohol, the lowest molar mass representative of this oligomer series (n=0), was used as reference compound in this case. Finally, the peak set (d) corresponds to the α,ω-dibenzyloxy oligomer series (functionality fraction 4). This functionality fraction does not appear as a single peak but shows a separation according to molar mass. The last peak of the chromatogram of sample A at a retention time of 36.28 min can be assigned to formaldehyde dibenzyl acetal, which is the lowest molar mass representative (n=0) of functionality fraction 4.

The chromatographic behaviour of functionality fraction 4 can be explained considering the pore size of the stationary phase. Since the average pore diameter (300 Å) is larger than the size of the solvated α,ω-dibenzyloxy PTOs, only one benzyloxy group can interact with the pore walls of the stationary phase. The second benzyloxy group contributes to the specific solvation of the PTO chain causing a slight deviation of the critical mobile phase composition. Accordingly, this functionality fraction is not exactly at the critical point but shows an oligomer separation.

Evaluation

The mass portions of the four functionality fractions can be determined by calibrating the peak areas with reference samples. Reference samples are obtained by preparatively separating samples A and B. For the determination of the molar mass distributions, the functionality fractions are analysed by SEC (see for example the chromatograms of fractions from sample A in Fig. 6.20).

The individual peaks in the size exclusion chromatograms are assigned by calibration with the respective first representative of each functionality fraction (n=0), i.e. diethylene glycol for 2, benzyl alcohol for 1 and formaldehyde dibenzyl acetal for 4. With this assignment, oligomer calibration curves can be obtained for each functionality fraction (see Fig. 6.21), which are subsequently used for the calculation of the molar mass distributions of the functionality fractions.

By knowing the mass portions and the molar mass distributions of the functionality fractions, the total molar mass of samples A and B can be calculated and correlated to data obtained independently by other methods.

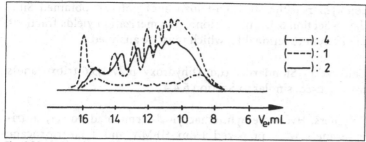

Fig. 6.20. SEC analysis of functionality fractions of sample A, RI detection; stationary phase: styrene – divinylbenzene; mobile phase: tetrahydrofuran (Reprinted from Ref. [18] with permission of Hüthig & Wepf Publishers, Switzerland)

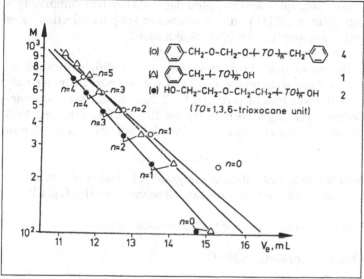

Fig. 6.21. SEC calibration curves for the functionality fractions 1, 2 and 4 (Reprinted from Ref. [18] with permission of Hüthig & Wepf Publishers, Switzerland)

6.4.5 Analysis of Hydroxyethylmethacryloyl-terminated Poly(1,3,6-trioxocane)s

Macromonomers are important precursors for the preparation of   Aim
block and graft copolymers. They can be synthesized by cationic
ring opening polymerization of oxygen-containing heterocycles.
In the presence of 2-hydroxyethyl methacrylate (HEMA), 1,3,6-
trioxocane can be polymerized, yielding hydroxyethylmethacryloyl-terminated poly(1,3,6-trioxocane). The methacrylic group
then may be copolymerized with other methacrylates, and graft

copolymers with poly(1,3,6-trioxocane) grafts are obtained. Similar to Section 6.3.4, the cationic polymerization yields fractions of different functionality, which must be analysed.

Materials

Calibration Standards. α,ω-Dihydroxy poly(1,3,6-trioxocane)s may be used, similar to Section 6.4.4.

Polymers. Hydroxyethylmethacryloyl-terminated poly(1,3,6-triox-ocane)s were prepared from HEMA and 1,3,6-trioxocane according to a procedure decribed in [19]. The structural parameters of the products are summarized in Table 6.4.

Equipment

Chromatographic System. Modular HPLC system comprising a conventional HPLC pump, a Rheodyne six-port injection valve and any conventional HPLC column oven.

Columns. Critical chromatography: C_{18}-modified silica gel, 10 mm average particle size and 300 Å average pore diameter. Column size was 250x4.6 mm I.D.. SEC: Two column-set of styrene-divinylbenzene gel with exclusion limits of 12 000 and 4000 g/mol, 10 μm average particle size. Column size was 300x8 mm I.D.

Mobile Phase. Critical chromatography: Mixtures of acetonitrile and water, SEC: tetrahydrofuran, all solvents are HPLC grade.

Detectors. Waters 410 differential refractometer.

Column Temperature. 25 °C

Sample Concentration. 3 mg/mL for critical chromatography and 0.5 mg/mL for SEC. All samples are dissolved in the mobile phase.

Injection Volume. 20 μL

Table 6.4. Structural parameters of the polymers

Sample	Molar Ratio TO/HEMA	OH-number (mg KOH/g)	I_2-number (g I_2/100 g)
A1	1/1	212	101
A5	10/1	45	21
B1	2/1	143	67

The OH-number and the I_2-number are quantitative measures for the number of OH-groups and double bonds per molecule, respectively.

Since in the present application polymer samples with a PTO backbone are to be investigated, the critical conditions described in Section 6.4.4 can be used. The critical point of adsorption for PTO corresponds to a mobile phase composition of acetonitrile-water 49.5:50.5% by volume.

Preparatory Investigations

With the use of chromatographic conditions corresponding to the critical point of adsorption of poly(1,3,6-trioxocane), samples A1 and A5 are separated with respect to the end groups (see Fig. 6.22). Similar to the previous investigations, four elution peaks are obtained, corresponding to the functionality fractions 1-4. The identification of the elution peaks was carried out by comparison with the elution behaviour of the lowest molar mass representative (n=0) of each oligomer series.

Separations

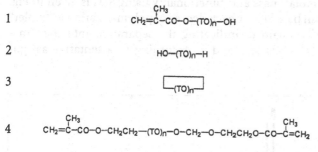

1      $CH_2=\overset{\overset{\displaystyle CH_3}{|}}{C}-CO-O-(TO)_n-OH$

2      $HO-(TO)_n-H$

3      $\boxed{-(TO)_n-}$

4      $CH_2=\overset{\overset{\displaystyle CH_3}{|}}{C}-CO-O-CH_2CH_2-(TO)_n-O-CH_2-O-CH_2CH_2O-CO-\overset{\overset{\displaystyle CH_3}{|}}{C}=CH_2$

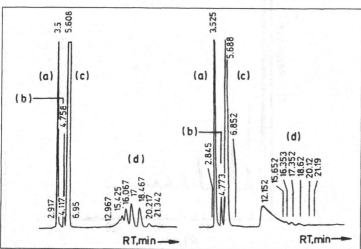

Fig. 6.22. Functionality chromatograms of samples A1 and A5; stationary phase: RP-18; mobile phase: acetonitrile-water 49.5:50.5% by volume (Reprinted from Ref. [19] with permission of Hüthig & Wepf Publishers, Switzerland)

Accordingly, peak (a) corresponds to the α,ω-dihydroxy oligomers (functionality fraction 2), whereas peak (b) is due to the cyclic oligomers (functionality fraction 3). Peak (c) having the highest intensity can be assigned to functionality fraction 1, and the peak series (d) corresponds to the functionality fraction 4.

Evaluation

The mass portions of the four functionality fractions can be determined by calibrating the peak areas via preparative separations. To describe the HEMA-terminated poly(1,3,6-trioxocane)s completely, it is necessary to determine the functionality type distribution as well as the molar mass distribution of each functionality fraction and the average molar mass of the whole sample. Similar to previous examples in this chapter this may be achieved by analysing the functionality fractions by SEC (see Section 6.4.4).

Supercritical fluid chromatography (SFC) has been shown to be a unique method for the simultaneous determination of FTD and MMD [20]. The separation of sample B1 into oligomers of different molar mass and functionality using SFC is given in Fig. 6.23. As can be seen, for each degree of polymerization a "triplet" structure is obtained, indicating the separation into the functionality fractions 1, 2 and 4. According to a tentative assign-

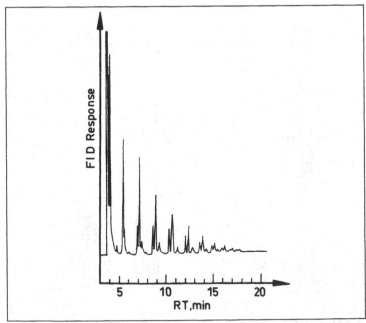

Fig. 6.23. Supercritical fluid chromatogram of sample B1 (Reprinted from Ref. [19] with permission of Hüthig & Wepf Publishers, Switzerland)

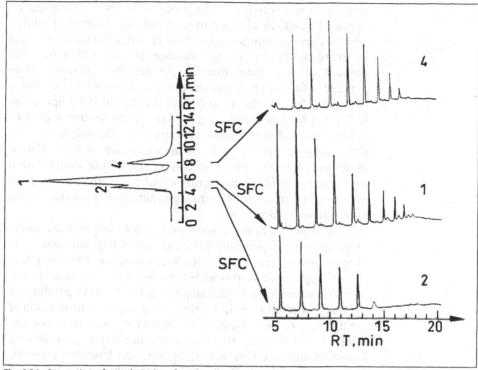

Fig. 6.24. Separation of sample B1 into functionality fractions 1, 2 and 4 and their SFC analysis (Reprinted from Ref. [19] with permission of Hüthig & Wepf Publishers, Switzerland)

ment, the separation with respect to functionality occured in the order *4-1-2*.

In order to support this interpretation, sample B1 was separated into fractions of single functionality by preparative critical chromatography. The resulting fractions were then subjected to SFC to give the oligomer distribution (see Fig. 6.24). By comparison of the elution times of the oligomer peaks, the tentative assignment was proven to be valid. Each single peak could be assigned to a certain degree of polymerization and functionality. Based on this assignment the molar mass can be calculated for each functionality fraction and for the total sample B1.

6.5 Separation of Block Copolymers

6.5.1 Principles and Limitations

Copolymers are complex macromolecular systems that are formed when two or more monomers of different chemical structures react in a polymerization process. As a result products are obtained which, in addition to the MMD, are characterized by a certain chemical composition. This chemical composition may be expressed by an average number, characterizing the total amount of each monomer in the reaction product. More detailed information, however, is obtained when the chemical heterogeneity (or the distribution in chemical composition, CCD) may be determined, characterizing the sequence distribution of the different monomer units along the polymer chain. Depending on the sequence of incorporation of the different monomers into the polymer chain, alternating and random copolymers, or block and graft copolymers are obtained. In addition, homopolymers of the different monomers are formed as unwanted by-products.

Block copolymers are usually prepared by sequential anionic or group transfer polymerization. In a first step monomer A is polymerized to form the first block with reactive chain ends. In a second step monomer B is added to react with the chain ends of block A. Accordingly, in this step block B is built. Depending on the reaction conditions, in particular the sequence of addition of the different monomers, diblock, triblock or multiblock copolymers can be formed. In some cases random copolymers or homopolymers are obtained along with the block copolymers. For a complete description of a block copolymer the following structural parameters must be determined:

– molar mass distribution of the block copolymer
– amount and molar mass of homopolymers
– amount and molar mass of random copolymers
– size of the single blocks of the copolymer.

The analysis of block copolymers by SEC is described in detail in a review by Kilz and Gores [48]. The analysis of block copolymers by gradient HPLC and chromatographic cross-fractionation has been extensively discussed ref. [21]. In brief, the complex molar mass-chemical composition distribution of copolymers requires separation in more than one direction. The classical approach is based upon the dependence of copolymer solubility on composition and chain length. A solvent/non-solvent combination fractionating solely by molar mass would be appropriate for the evaluation of the MMD, another one separating by chemical composition would be suited for determining the CCD of the copolymer. In general, fractionation is influenced by the molar mass *and* chemical composition, and the direction of separation is determined by the experimental conditions, in particular by the solvent/non-solvent combination chosen.

Chromatographic retention may be directed by entropic and enthalpic interactions (see Chapter 2). In particular, enthalpic interactions of the solute molecules and the packing may be used for the separation of copolymers with respect to chemical composition. Very specific precipitation/redissolution processes are able to promote separation with respect to chemical composition. With the use of solvent mixtures as the mobile phase, the precipitation/redissolution equilibria may be adjusted, and changing the composition of the mobile phase during the elution process the solubility of the sample fractions may be changed. Thus, using gradient elution techniques, the polymer sample may be fractionated with respect to solubility and, accordingly, with respect to chemical composition [22–24].

Another approach to the chromatographic characterization of block copolymers is the concept of "invisibility", which assumes that chromatographic conditions exist, under which heteropolymers may be separated according to the size of one of the components, because the other component is chromatographically "invisible", i.e., does not contribute to retention. This concept experimentally relates to liquid chromatography at the critical point of adsorption, which was described in detail in Section 6.1. The usefulness of this technique for the determination of FTD of functional homopolymers was demonstrated in Section 6.4.

The application of liquid chromatography at the critical point of adsorption to block copolymers is based on the consideration that Gibbs free energy ΔG_{AB} of a block copolymer A_nB_m is the sum of the contributions of block A and block B, ΔG_A and ΔG_B respectively.

$$\Delta G_{AB} = \Sigma n_A \Delta G_A + n_B \Delta G_B + \chi_{AB} \qquad (6.9)$$

where χ_{AB} describes the interactions between blocks A and B. Assuming no specific interactions between A and B ($\chi_{AB}=0$), the change in Gibbs free energy is only a function of the contributions of A and B.

$$\Delta G_{AB} = \Sigma n_A \Delta G_A + n_B \Delta G_B \qquad (6.10)$$

By the use of chromatographic conditions, corresponding to the critical point of homopolymer A, block A in the block copolymer will be chromatographically invisible, and the block copolymer will be eluted solely with respect to block B.

$$\Delta G_A = 0$$

$$\Delta G_{AB} = \Sigma n_B \Delta G_B \qquad (6.11\text{–}6.13)$$

$$K_d^{AB} = K_d^{B}$$

At the critical point of homopolymer B, block B will be chromatographically invisible, and the block copolymer will be eluted solely with respect to block A.

$$\Delta G_B=0$$

$$\Delta G_{AB}=\Sigma n_A \Delta G_A \qquad (6.14\text{–}6.16)$$

$$K_d^{AB}=K_d^A$$

The first evidence for the validity of this approach was given by Gankina et al. [25] for the analysis of block copolymers by thin layer chromatography. Column liquid chromatography was used by Zimina et al. [26] for the analysis of poly(styrene-block-methyl methacrylate) and poly(styrene-block-t.-butyl methacrylate). However, the critical conditions were established only for the polar part of the block copolymers, i.e., PMMA and PtBMA, respectively. Thus, only the polystyrene block was analysed.

Depending on the polarity of blocks A and B in the block copolymer A_nB_m, and the polarity of the stationary phase, different chromatographic situations can be encountered, see Fig. 6.25, where ε indicates the interaction energy and ε_c corresponds to

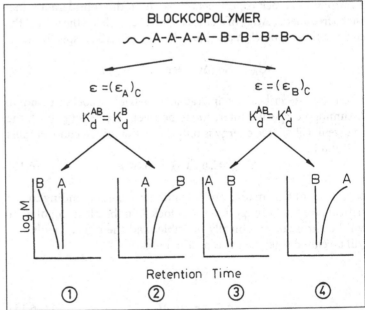

Fig. 6.25. Schematic representation of different chromatographic situations in liquid chromatography at the critical point of adsorption of block copolymers (for explanation of 1–4 see text)

the interaction energy at the critical point of adsorption. For example, at the critical point of homopolymer A ($\varepsilon = (\varepsilon_A)_c$), homopolymer B may be separated either in the SEC mode (1) or the adsorption mode (2). The same is true for the critical point of B ($\varepsilon = (\varepsilon_B)_c$), where A may be eluted according to (3) or (4). Preferable, of course, are cases 1 and 3, whereas in cases 2 and 4 for high molar mass polymers irreversible adsorption may be encountered.

The interrelations between the polarity of a specific block in a block copolymer and the polarity of the stationary phase to be chosen is summarized in Fig. 6.26. Let us consider that the polarity of block A in the block copolymer is higher than the polarity of block B (polarity A>B). In this case, chromatographic behaviour according to (1) is achieved, when normal-phase silica gel is used as the stationary phase (silica gel separates in the order of *increasing* polarity). Separation according to (3) is obtained on a reversed-phase column, such as RP-8 or RP-18 (separation in the order of *decreasing* polarity). If now the polarity of A is lower than the polarity of B (polarity A<B), on a normal-phase silica gel separation according to cases 2 and 3 may be obtained, whereas on a reversed phase cases 1 or 4 are operating. Using these considerations, the appropriate stationary phase may be preselected for each specific case of separation.

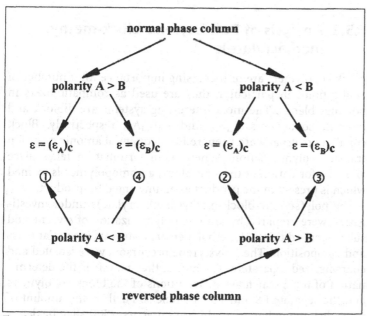

Fig. 6.26. Chromatographic behaviour of blocks of different polarity on normal and reversed stationary phases

Of course, there are several limitations of the method with respect to the analysis of block copolymers. The first limitation is the solubility of the different blocks of the block copolymer in the mobile phase. As the blocks usually have different polarities, the solubility behaviour of the corresponding homopolymers may differ remarkably. If the difference in polarity is to large it might be impossible to find common solvents, i.e., the solvent for one block in all cases is a precipitant for the other block. Therefore, in preliminary experiments (cloud point titration) the solubility of the corresponding homopolymers in the mobile phase must be investigated to make sure that a real solution of the block copolymer is obtained, and the formation of aggregates, associates, or micelles does not occur. A second limitation of the method is the selectivity of the stationary phase. Operating at the critical point of one block B, the block copolymer shall elute, say, in the SEC mode with respect to the other block A (see cases 1 and 3 in Fig. 6.25). Quantitative determinations, e.g. the determination of the MMD of block A, can be carried out, when the corresponding homopolymer A exhibits ideal SEC behaviour. This can be tested by running the corresponding homopolymer A in a thermodynamically good mobile phase and in the mobile phase corresponding to the critical point of B. If the molar mass vs retention time behaviour in both mobile phases coincides, proper quantification can be carried out.

6.5.2 Analysis of Poly(styrene-block-methyl methacrylate)s

Aim

Block copolymers are of increasing importance for a number of applications. In particular, they are used as compatibilizers in polymer blends. The most interesting systems are diblock and triblock copolymers A_nB_m and $A_nB_mA_n$, respectively. Block copolymers usually are prepared by sequential anionic or group transfer polymerization. When chain termination takes place after the formation of the first block, a homopolymer is formed which is present in the product as an unwanted by-product.

The poly(styrene-block-methyl methacrylate)s under investigation were prepared by anionic polymerization of styrene and subsequent addition of methyl methacrylate, varying molar mass and composition. The polystyrene precursors were isolated and characterized separately. The aim of the analysis is the determination of the molar mass distributions of the block copolymers and the separate PS and PMMA blocks. Further, the amount of possible homopolymer (in this case precursor PS) shall be determined.

Calibration Standards. Narrow-disperse polystyrene (PS) and   Materials
poly(methyl methacrylate) (PMMA) in the molar mass range of
1 000 to 300 000 g/mol.

Polymers. Anionically polymerized poly(styrene-block-methyl
methacrylate)s of different molar masses and compositions and
the corresponding PS precursor polymers. The average structure
of the samples used in the present study, as supplied by the manufacturer (Y. Gallot, Institut Charles Sadron, Strasbourg, France)
is summarized in Table 6.5.

Chromatographic System. Modular HPLC system comprising a   Equipment
Waters Model 510 pump, a Rheodyne six-port injection valve and
a Waters column oven.

Columns. Critical chromatography: Nucleosil Si-100 (Macherey-
Nagel), 5 µm average particle size and 100 Å average pore size,
with a column size of 200x4 mm I.D., or a two column-set of
LiChrospher Si-300 and Si-1000 (Merck), 10 µm average particle
size and 300 Å and 1000 Å average pore size, with column sizes of
250x4.6 mm I.D., or a two column-set of Nucleosil RP-18
(Macherey-Nagel), 7 µm average particle size and 300 Å and 1000
Å average pore size, with column sizes of 250x4.6 mm I.D.; SEC:
Six column-set of Ultrastyragel (Waters) $10^6 + 2x10^5 + 2x10^4 + 10^3$
Å, 10 µm average particle size and column sizes of 300x8 mm I.D.

Mobile Phase. Critical chromatography: mixtures of methyl ethyl
ketone and cyclohexane or tetrahydrofuran and acetonitrile,
SEC: tetrahydrofuran. All solvents are HPLC grade.

Detectors. Waters 410 differential refractometer and Knauer
fixed wavelength UV /vis detector at a wavelength of 280 nm.

Table 6.5. Average structure of the block copolymers (B1-B3) and
the corresponding PS precursors (P1-P3)

Sample	St/MMA (molar%)	M_w (SEC) (g/mol)
B1	67/33	165 000
P1	100/0	110 000
B2	51/49	182 000
P2	100/0	93 000
B3	29/71	188 000
P3	100/0	55 000

Column Temperature. 25 °C

Sample Concentration. 1–5 mg/mL for critical chromatography and 0.5 mg/mL for SEC. All samples are dissolved in the mobile phase.

Injection Volume. 20–50 μL

Preparatory Investigations

As was explained in Section 6.5.1, the single blocks of the block copolymers can be analysed using critical chromatographic conditions for the respective blocks. Accordingly, for the poly(styrene-block-methyl methacrylate)s the critical points of adsorption of PS and PMMA must be found. Following the discussion on the elution behaviour as a function of column polarity, see Fig. 6.25, different stationary phases must be selected for establishing the critical points of PS and PMMA, respectively.

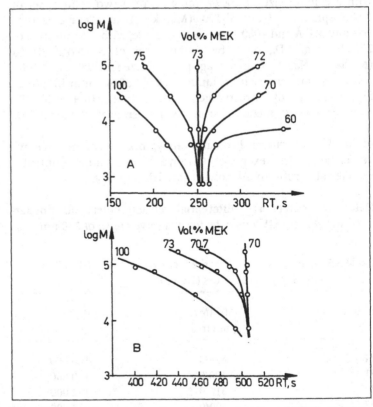

Fig. 6.27. Critical diagrams of molar mass vs retention time of PMMA; stationary phase: Nucleosil Si-100 **(A)** or LiChrospher Si- 300+Si-1000 **(B)**; mobile phase: methyl ethyl ketone- cyclohexane (Reprinted from Ref. [27] with kind permission from Elsevier Science Ltd., UK)

Since the PMMA block is the more polar block in the block copolymers, a polar (silica gel) column is chosen for establishing its critical point. According to case (1) in Fig. 6.25, the PS block is then eluted in the SEC mode. The behaviour of PMMA of different molar masses on silica gel Si-100 in eluents, comprising methyl ethyl ketone and cyclohexane is given in Fig. 6.27A [27].

The figure indicates that at concentrations of methylethylketone >73% by volume in the mobile phase, separation is predominantly driven by entropic effects and the SEC mode is operating. In contrast, the LAC mode with predominantly enthalpic interactions is operating at concentrations of methyl ethyl ketone <73% by volume in the eluent. The critical point of adsorption of PMMA is obtained at an eluent composition of methyl ethyl ketone-cyclohexane 73:27% by volume. At this point, all PMMA samples, regardless of their molar mass elute at one retention time. This, by definition, indicates that the PMMA polymer chain behaves "chromatographically invisible", i.e., does not contribute to retention. Accordingly, using these chromatographic conditions, block copolymers of PMMA and PS can be analysed with respect to the PS block.

Depending on the size of the macromolecules under investigation, similar to conventional SEC, the pore size of the packing must be adjusted to the desired molar mass range. Thus, for higher molar mass samples, the investigations must be carried out on column sets with larger pores, see Fig. 6.27B for a two column-set of LiChrospher Si-300 + Si-1000. For this column set critical conditions were found to be operating at an eluent composition of methyl ethyl ketone-cyclohexane 70:30% by volume.

Analogous to the analysis of the PS block, the PMMA block of the block copolymers can be characterized at chromatographic conditions, corresponding to the critical point of PS. Formally, the critical point of adsorption of PS can be established on different stationary phases. By the use of a silica gel column, at the critical point of PS the PMMA would be eluted in the adsorption mode (see case 4 in Fig. 6.25). As retention time in the LAC mode is exponentially related to molar mass, irreversible adsorption would be likely to occur for higher molar mass samples. Therefore, a reversed-stationary phase is more feasible for establishing chromatographic conditions, corresponding to case 3 in Fig. 6.25. The retention behaviour of PS on a two-column set of Nucleosil RP-18 with pore sizes of 300 Å and 1000 Å, and a mobile phase comprising tetrahydrofuran and acetonitrile is shown in Fig. 6.28 [28]. The critical point of adsorption corresponds to a mobile phase composition of tetrahydrofuran-acetonitrile 49:51% by volume. With these chromatographic conditions, the PMMA block can selectively be analysed regardless of the PS block.

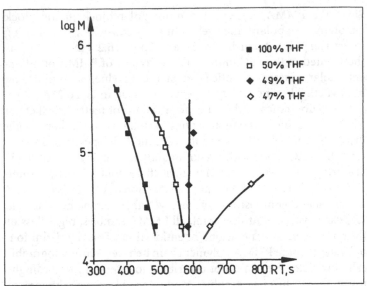

Fig. 6.28. Critical diagram of molar mass vs retention time of polystyrene; stationary phase: Nucleosil RP-18 300 Å + 1000 Å; mobile phase: tetrahydrofuran-acetonitrile (Reprinted from Ref. [28] with kind permission from Elsevier Science Ltd., UK)

Separations

In a first set of experiments, the block copolymers and the corresponding PS precursor blocks are investigated in the ideal SEC mode for both components. The ideal SEC behaviour for PS and PMMA is obtained on conventional polystyrene columns (see Chapter 4 on SEC) using tetrahydrofuran as the mobile phase. The chromatograms obtained on a six-column set clearly indicate that the samples are separated with respect to their total molar masses (see Fig. 6.29A). The PS precursors elute at higher retention times compared to the corresponding block copolymers, as expected due to their lower molar masses. The quantitative determination of the molar masses agrees well with the numbers given by the manufacturer.

The second set of experiments is carried out at chromatographic conditions corresponding to the critical point of adsorption of PMMA. In this case, the PMMA blocks are supposed to behave "chromatographically invisible" and the block copolymers are expected to elute with respect to the PS blocks.

The chromatograms of the block copolymers B1-B3 and the corresponding PS precursors P1-P3 at the critical point of adsorption with respect to PMMA are given in Fig. 6.29B. As can be seen, the chromatograms of the copolymers and the precursors have similar shapes and retention ranges. In contrast to ideal SEC for both blocks, where the chromatograms of B1-B3 and

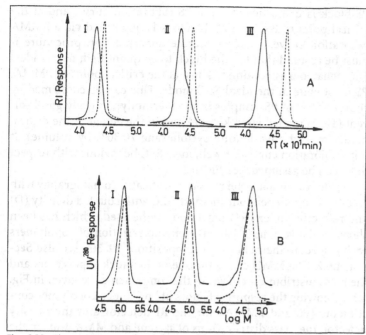

Fig. 6.29. Chromatograms of poly(styrene-block-methyl methacrylate)s (—) and their polystyrene precursors (– – –) in the SEC mode for both blocks **(A)** and at the critical point for the PMMA block **(B)**; stationary phase: Ultrastyragel **(A)** or silica gel **(B)**; mobile phase: tetrahydrofuran **(A)** or methyl ethyl ketone-cyclohexane 70:30% by volume; samples: I-B1, P1, II-B2, P2, III-B3, P3 (Reprinted from Ref. [27] with kind permission from Elsevier Science Ltd., UK)

P1-P3 are well separated, in the case of critical chromatography a complete overlay of the block copolymers and the PS precursors is obtained. These data indicate that at the critical point of PMMA the block copolymers behave like the corresponding PS precursors. Since the block copolymers elute selectively with respect to the PS blocks in the SEC mode, a conventional SEC calibration curve for PS can be used for quantifying the PS blocks.

In addition to the selective separation mode, a selective detection would improve the accuracy of the experiments. The PS blocks can selectively be detected by using a UV detector at a wavelength of 280 nm, where PMMA does not absorb light.

Similar to the PS blocks, the PMMA blocks can selectively be analysed using chromatographic conditions, corresponding to the critical point of PS. These can be established on a stationary phase of RP-18, 300 Å + 1000 Å, using a mobile phase of tetrahydrofuran-acetonitrile 49:51% by volume.

The quantitative determination of the molar masses of the PS and PMMA blocks can be carried out using conventional SEC calibration procedures. Thus, at the critical point of PMMA, the

Evaluation

PS block is quantified using a PS calibration curve, and at the critical point of PS, the PMMA block is quantified via a PMMA calibration curve. For an accurate quantification procedure it must be assured that for the block to be quantified, nearly ideal SEC behaviour is obtained. Thus, at the critical point of PMMA, PS must elute in the ideal SEC mode. This can be confirmed by running a set of PS samples in the thermodynamically good solvent (100% methyl ethyl ketone) and in the eluent of the critical mode (methyl ethyl ketone-cyclohexane 70:30% by volume). If both calibration curves fit well, ideal SEC behaviour with respect to PS can be assumed, see Fig. 6.30.

In order to compare the results of critical chromatography with results of an independent method, SEC with coupled density (D) and refractive index (RI) detection can be used, which has been shown to be very useful for the characterization of copolymers with respect to their chemical composition [29, 30], see also Section 4.6.2. The MMD curve for one of the block copolymers and the mass distribution curves of the components are given in Fig. 6.31. Knowing the response factors of both block copolymer constituents (PS and PMMA) for the RI detector and the density detector, the mass distributions of styrene and MMA units in the block copolymer can be determined at each point of the chro-

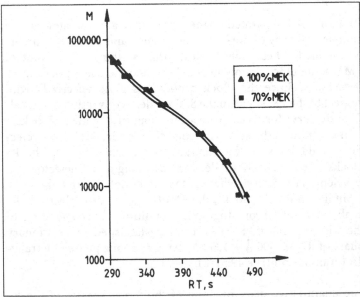

Fig. 6.30. Calibration curves molar mass vs retention time of polystyrene; stationary phase: LiChrospher Si-300 + Si-1000; mobile phase: methyl ethyl ketone or methylethylketone-cyclohexane 70:30% by volume

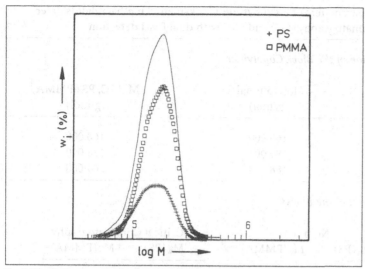

Fig. 6.31. Molar mass distribution of the block copolymer B3 and the mass distributions of the components by SEC with coupled D-RI detection; stationary phase: styragel; mobile phase: chloroform (Reprinted from Ref. [28] with kind permission from Elsevier Science Ltd., UK)

matogram. Then, knowing the chemical composition of the sample at each point of the chromatogram, a copolymer calibration curve can be constructed from the calibration curves of the homopolymers, and the MMD of the block copolymer can be determined.

From these data the overall chemical composition may be calculated. The results obtained by SEC with D-RI detection agree very well with the results of critical chromatography (see Table 6.6). This is a clear documentation for the usefullness of critical chromatography in block copolymer analysis. The molar masses Mw(PS) and Mw(PMMA) are determined by chromatography at the critical point of PMMA and PS, respectively. From these values the molar masses of the block copolymers are calculated by summing Mw(PS) + Mw(PMMA). The chemical composition is calculated from the ratio Mw(PS)/Mw(PMMA).

Table 6.6. Molecular parameters of poly(styrene-block-methyl methacrylate)s determined by critical chromatography (CC) and SEC with dual D-RI detection

Molar Mass Distribution of the Block Copolymers

Sample	M_w(Nominal) (g/mol)	M_w(CC, PS+PMMA) (g/mol)
B1	165 000	168 000
B2	182 000	188 000
B3	188 000	204 000

Molar Mass Distribution of the Blocks

Sample	Nominal		Critical Chromatography	
	M_w(PS)	M_w(PMMA)	M_w(PS)	M_w(PMMA)
B1	110 000	55 000	119 000	49 000
B2	93 000	89 000	91 000	97 000
B3	55 000	133 000	61 000	143 000

Chemical Composition of the Block Copolymers (St/MMA in mol-%)

Sample	Nominal	CC	SEC (D-RI)
B1	67/33	71/29	69/31
B2	51/49	48/52	53/47
B3	29/71	30/70	30/70

6.5.3 Analysis of Poly(decyl methacrylate-block-methyl methacrylate)s

Aim

The combination of polymethacrylates of different polarities is interesting with respect to compatibilization of polymer blends and micelle forming polymer systems. In particular, diblock copolymers of polydecyl methacrylate (PDMA) and polymethyl methacrylate (PMMA) fit this purpose. These block copolymers can be prepared by group transfer polymerization, forming the PDMA block in the first step, and then adding methyl methacrylate to form the PMMA block. In some cases, the preformed PDMA block undergoes secondary reactions, resulting in the desactivation of the reactive terminal group. In these cases, PDMA homopolymer will be present in the final product.

Table 6.7. Average structure of the poly(decyl methacrylate-block-methyl methacrylate)s

Sample	DMA/MMA (molar%)	M_w (SEC) (g/mol)
A1	25/75	59 000
A3	25/75	121 000
A4	50/50	149 000
A5	50/50	78 000

Calibration Standards. Narrow-disperse PMMA and a number of PDMA in the molar mass range of 10 000 to 200 000 g/mol.   Materials

Polymers. Group transfer-polymerized poly(decyl methacrylate-block-methyl methacrylate)s of different molar masses and compositions. The samples used for the present investigation are laboratory samples of Röhm Chemische Fabrik GmbH, Darmstadt, Germany.

Chromatographic System. Modular HPLC system comprising a   Equipment
Waters Model 510 pump, a Rheodyne six-port injection valve and a Waters column oven.

Columns. Critical chromatography: two column-set of LiChrospher Si-300 and Si-1000 (Merck), 10 µm average particle size and 300 Å and 1000 Å average pore size, with column sizes of 250x4.6 mm I.D., or a two column-set of Nucleosil RP-18 (Macherey-Nagel), 7 µm average particle size and 300 Å and 1000 Å average pore size, with column sizes of 250x4.6 mm I.D.

Mobile Phase. Critical chromatography: mixtures of methyl ethyl ketone and cyclohexane or chloroform and ethanol.

Detectors. Waters 410 differential refractometer.

Column Temperature. 25 °C.

Sample Concentration. 1–5 mg/mL for critical chromatography and 0.5 mg/mL for SEC. All samples are dissolved in the mobile phase.

Injection Volume. 20–50 µL.

Preparatory Investigations

Similar to the previous experiments, the critical points of adsorption for PMMA and PDMA must be determined. The critical point of PMMA was found in Section 6.5.2 to correspond to a mobile phase composition of methyl ethyl ketone-cyclohexane 70:30% by volume and a stationary phase of LiChrospher 300+1000 Å.

The critical conditions for PDMA are established on a nonpolar stationary phase. By the use of the column combination Nucleosil RP-18 300 Å+ 1000 Å, and considering the solubility behaviour of PDMA, an eluent comprising chloroform as the good solvent and ethanol as the poor solvent can be used. For this phase system the critical point of adsorption corresponds to a mobile phase composition of chloroform-ethanol 61:39% by volume, see Fig. 6.32 [31].

Separations

Before determining the PDMA and PMMA blocks by critical chromatography, the total molar masses of the block copolymers can be determined by SEC. This is easily done by using a mobile phase of 100% methylethylketone as the mobile phase and LiChrospher Si-300+1000 as the packing, thus substanciating a size exclusion separation mechanism with respect to both blocks, see Fig. 6.33. Then, by changing the composition of the mobile phase to methylethylketone-cyclohexane 70:30% by volume, the critical mode of PMMA is established. Under these conditions, the block copolymers are eluted exclusively with respect to the

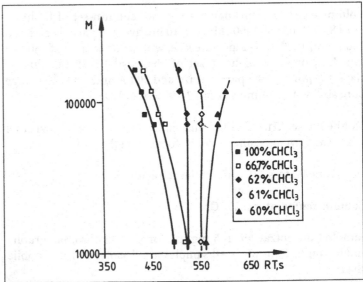

Fig. 6.32. Critical diagram molar mass vs retention time of PDMA, stationary phase: Nucleosil RP-18 300+1000 Å, mobile phase: chloroform-ethanol

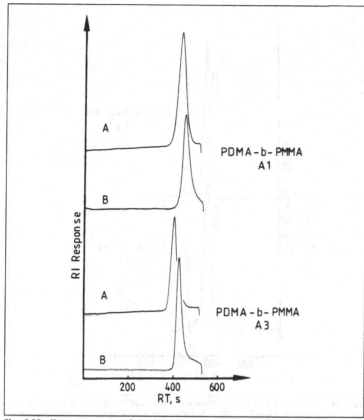

Fig. 6.33. Chromatograms of the block copolymers in different chromatographic modes; stationary phase: LiChrospher Si- 300+1000; mobile phase: 100% methyl ethyl ketone **(A)** or methylethylketone-cyclohexane 70:30% by volume **(B)**

PDMA blocks. At the critical point of PMMA, PDMA is eluted in the size exclusion mode, see Fig. 6.33.

The present elution curves exhibit a single, symmetrical elution peak for each sample, thus indicating that PMMA homopolymer is not present in the samples.

Changing now the chromatographic conditions to the critical point of PDMA, the PMMA blocks can be analysed selectively. The chromatograms of these separations are presented in Fig. 6.34, the packing being Nucleosil RP-18 and the eluent being chloroform-ethanol. The chromatograms show, in addition to the main elution peaks, small elution peaks at retention times of 540 s, corresponding to the critical point of PDMA. The main elution peaks are caused by the block copolymers, which are selectively eluted with respect to the PMMA blocks. In contrast, the

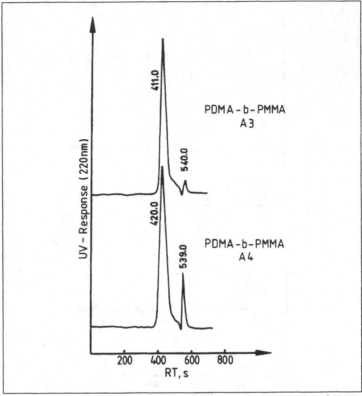

Fig. 6.34. Chromatograms of the block copolymers at the critical point of PDMA; stationary phase: Nucleosil RP-18 300+1000 Å; mobile phase: chloroform-ethanol 61:39% by volume

small elution peaks are not caused by the block copolymers, but by PDMA homopolymer present in the samples.

Evaluation

The quantitative determination of the molar masses of the PDMA and PMMA blocks can be carried out using conventional SEC calibration procedures. Under critical conditions of PMMA, the PDMA blocks are determined using a PDMA calibration curve, and the PMMA blocks are quantified under critical conditions of PDMA via a PMMA calibration curve [31, 32].

For an additional check of the validity of the results obtained by critical chromatography, SEC with dual D-RI detection can be carried out, compare Section 6.5.2. The molar mass distribution curves and the mass distribution curves of the components are shown in Fig. 6.35. In all cases the mass distribution of PDMA is more heterogeneous than the PMMA mass distribution. At the

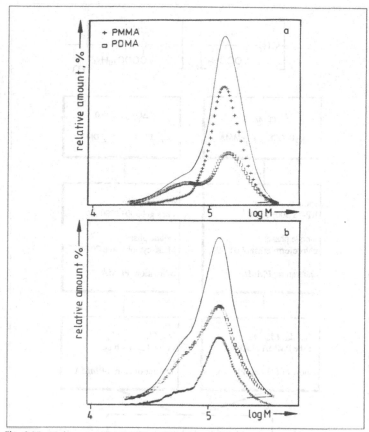

Fig. 6.35. Molar mass distributions of the block copolymers A3 (a) and A4 (b) and mass distributions of the monomers from SEC (D-RI); stationary phase: styragel; mobile phase: chloroform

lower molar mass end of the distribution curves a certain amount of PDMA homopolymer seems to appear, which is in full agreement with Fig. 6.34. Obviously, this PDMA homopolymer, which was formed in the first polymerization step, failed to continue the polymerization upon addition of the MMA monomer. A similar result was obtained by HPLC separations according to chemical composition [33].

Figure 6.34 shows that using chromatography at the critical point of adsorption it is not only possible to analyse the single blocks of the block copolymers but also traces of homopolymers, which might be present in the sample. The complete procedure of analysing poly(decyl methacrylate-block-methyl methacrylate)s by critical chromatography is summarized in the scheme, given in Fig. 6.36.

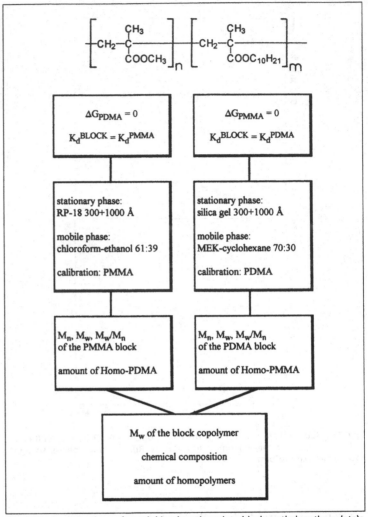

Fig. 6.36. Analysis scheme for poly(decyl methacrylate-block-methyl methacrylate)s

6.5.4 Analysis of Poly(ethylene oxide-block-propylene oxide-block-ethylene oxide)

According to Gorbunov et al. [34], triblock copolymers of the ABA' type may be analysed by liquid chromatography at the critical point of adsorption similar to the analysis of diblock copolymers. The two possible cases for this type of investigation are (a) the analysis with respect to the inner block B using the critical conditions of the outer blocks A and A' ($\Delta G_{A,A'}=0$), and (b) the analysis of the outer blocks A and A' using the critical conditions of the inner block B. It is particularly useful to carry out experiments at the critical point of A and A'. The separation occurs then with respect to the chain length of B, yielding fractions which are monodisperse with respect to B and polydisperse with respect to A and A'. These fractions can be analysed selectively with respect to the outer blocks A and A' in separate experiments.

$$\Delta G_{A,A'}=0 \tag{6.17}$$

$$\Delta G_{ABA'}=n_B\Delta G_B \tag{6.18}$$

$$K_d^{ABA'}=K_d^B \tag{6.19}$$

Random and block copolymers based on ethylene oxide (EO) and propylene oxide (PO) are important precursors of polyurethanes. Their detailed chemical structure, i.e., the chemical composition, block length, and molar mass of the individual blocks, may be decisive for the properties of the final product. For triblock copolymers $HO(EO)_n(PO)_m(EO)_nOH$ the detailed analysis relates to the determination of the total molar mass and the degrees of polymerization of the inner PPO block (m) and the outer PEO blocks (n).

Aim

Calibration Standards. Narrow disperse polyethylene oxide (PEO) in the molar mass range of 100–10 000 g/mol and a polypropylene glycol (PPG) of about 500 g/mol.

Materials

Polymers. Technical triblock copolymers with an average structure of $HO(EO)_n(PO)_m(EO)_nOH$. The sample under investigation was prepared at the the Central Institute of Organic Chemistry, Berlin, by anionic polymerization at 110 °C using potassium glycolate as initiator.

Chromatographic System. Modular HPLC system comprising a Waters Model 501 pump, a Rheodyne six-port injection valve and a Waters column oven.

Equipment

Columns. Critical chromatography: Nucleosil RP-18 (Macherey-Nagel), 5 μm average particle size and 100 Å average pore size, with a column size of 250x4 mm I.D., SEC: Five column-set of Ultrastyragel (Waters) 1000+2x500+2x100 Å, 10 μm average particle size and column sizes of 300x8 mm I.D.

Mobile Phase. Critical chromatography: mixtures of acetonitrile and water, SEC: tetrahydrofuran. All solvents are HPLC grade.

Detectors. Waters R-401 differential refractometer or 950/14 ACS mass detector.

Column Temperature. 25 °C.

Sample Concentration. 5–10 mg/mL for critical chromatography and 0.5 mg/mL for SEC. All samples are dissolved in the mobile phase.

Injection Volume. 20–50 μL.

Preparatory Investigations

For the selective separation of the block copolymer with respect to the inner PPO block, the experiments must be conducted using chromatographic conditions, corresponding to the critical point of the PEO outer blocks. As has been shown in Section 6.4.2, critical conditions for PEO can be established on a RP-18 stationary phase when a mobile phase of acetonitrile-water 42:58% by volume is used.

For determining the PEO blocks it is necessary to find the critical point for the inner PPO block. On a reversed-phase RP-18 we would then have the size exclusion mode for the outer PEO blocks. Unfortunately, it proved impossible to do this on the chosen stationary phase. Critical behaviour of PPO is obtained using an eluent of tetrahydrofuran-water with low water content. Under these conditions the unmodified sites of the packing, containing free silanol groups, become significant. Due to extremely strong interactions of PEO with these silanol groups, the chromatograms, even for the PEO homopolymer, are strongly distorted by elongated tails [5].

Separations

The separation of the triblock copolymer at the critical point of PEO is shown in Fig. 6.37. Under these conditions, the ethylene oxide blocks behave chromatographically "invisible" and retention of the block copolymer is solely directed by the propylene oxide block, yielding fractions of different degrees of polymerization m with respect to PPO.

The assignment of the peaks is based on comparison with the chromatogram of a PPG. The first and second peaks in the chromatogram at retention times of 244 and 259 s, respectively, could not be identified directly, whereas the third peak at a retention time of 275 s corresponds to m=1–2, m being the degree of polymerization with respect to PPO. The peak at 300 s corresponds to m=3, the peak at 326 s corresponds to m=4 and so on. Accordingly, every peak is uniform with respect to m but has a distribution in block length with respect to the PEO blocks (n).

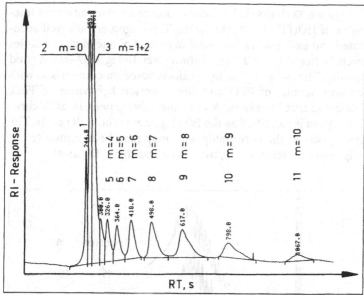

Fig. 6.37. Separation of a triblock copolymer $HO(EO)_n(PO)_m(EO)_nOH$ with respect to the PPO block; peak assignment indicates fraction and degree of polymerization of the PPO block (m); stationary phase: Nucleosil RP-18; mobile phase: acetonitrile-water 42:58% by volume.

In order to identify peaks 1 and 2, they are collected and subjected to mass spectrometry. With the use of matrix-assisted laser desorption/ionization, fraction 1 was found to contain only impurities. Fraction 2, however, showed a peak series, corresponding to the oligomer distribution of polyethylene glycol [35]. Accordingly, fraction 2 is due to the first member of the $HO(EO)_n(PO)_m(EO)_nOH$ oligomer series with m=0.

Owing to problems of solubility and specific interactions with the stationary phase, it was not possible to determine the outer PEO blocks at the critical point of PPO. This can be achieved, however, by subjecting the PO uniform fractions to a second chromatographic method. This method must separate the fractions with respect to the oligomer distribution of the PEO blocks, thus providing the oligomer distribution of these blocks.

In previous investigations on the two-dimensional separation of telechelic oligomers, it was demonstrated that supercritical fluid chromatography (SFC) is a useful technique for separating polyethers according to their oligomer distribution [20, 36]. With the use of highly efficient and selective capillary columns, oligomers may even be separated simultaneously according to the degree of polymerization and functionality.

Evaluation

Figure 6.38 shows the SFC chromatograms of a number of fractions of $HO(EO)_n(PO)_m(EO)_nOH$. The oligomers are well separated and each fraction is found to consist of one oligomer series each, indicating that the separation given in Fig. 6.37 was of good quality. The assigment of the peaks is based on comparison with chromatograms of PPO and the retention behaviour of PEO. Thus, the first intense peak after the solvent peak in each chromatogram is identified as the PO oligomer without EO units. The next peaks in the chromatograms may then be assigned to the oligomers containing one, two, three and so on, EO units.

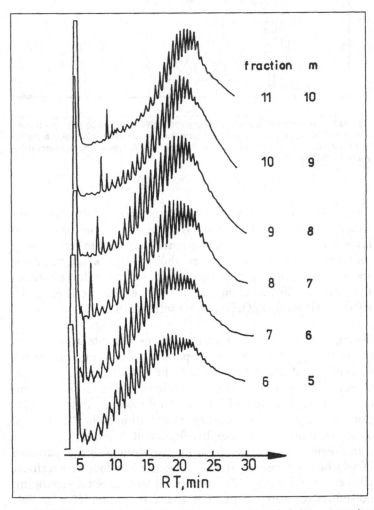

Fig. 6.38. SFC analysis of fractions 6–11 taken from the critical chromatography separation in Fig. 6.37; stationary phase: SB Biphenyl-30; mobile phase: carbon dioxide (Reprinted from Ref. [20], Copyright 1992 with kind permission from Elsevier Science Science – NL)

After assigning all peaks in the chromatograms to the corresponding chemical structure, i.e., determining m and n of $HO(EO)_n(PO)_m(EO)_nOH$ for each peak, quantification of the peaks can be carried out. This can be done via the area of each peak taking into account the corresponding response factor [20]. As a result, the molar mass average of each fraction is obtained. From these data and the relative concentrations of the fractions, taken from Fig. 6.37, the total molar mass of the block copolymer can be calculated.

For additional support of the results, the fractions taken from critical chromatography, are subjected to SEC (see Fig. 6.39). In this case the molar mass averages are calculated using a PEO calibration curve.

To summarize, combining different chromatographic methods it is possible to analyse triblock copolymers of PEO and PPO

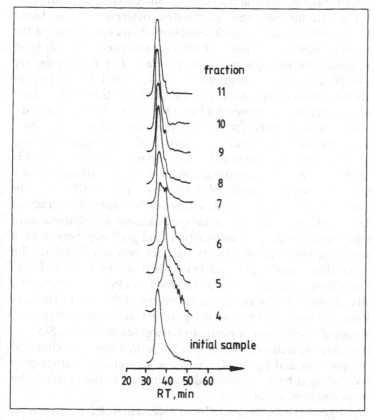

Fig. 6.39. SEC analysis of fractions 4–11 taken from the critical chromatography separation in Fig. 6.37; stationary phase: ultrastyragel; mobile phase: tetrahydrofuran (Reprinted from Ref. [20], Copyright 1992 with kind permission from Elsevier Science – NL)

with respect to the different blocks. By chromatography at the critical point of adsorption the block copolymer is separated with respect to the PPO oligomer distribution. The PEO oligomer distribution is then determined by SFC or SEC of propylene oxide uniform fractions.

6.6 Analysis of Polymer Blends

6.6.1 Principles and Limitations

Polymer blends, i.e., mixtures of two- or more polymeric components, are of increasing commercial importance for a number of applications. The advantage of polymer blends is the useful combination of the properties of the components without creating chemically new polymers. This approach in many cases is more feasible than developing new tailor-made polymer structures.

The identification and quantitative determination of blend components in most cases is complicated and depending on the chemical structure a variety of different analytical methods must be used. Spectroscopic methods, such as infrared spectroscopy [37–39] and nuclear magnetic resonance [40–42] may help to identify blend components. For the determination of the MMD of the components, however, often a separation step is required.

SEC has been used for the analysis of polymer blends by a number of authors [43–45]. However, as SEC separates according to hydrodynamic volume of the macromolecules, this method is limited to blends containing components of different molar mass. The separation of polymers according to differences in chemical composition and structure can be achieved by gradient elution HPLC. With this technique, random copolymers were separated according to composition, and graft copolymers were separated into a copolymer and a homopolymer fraction. The separation of mixtures of poly(methacrylates) by gradient HPLC has been successfully conducted by Mourey [46]. The effect of the alkoxy group on the retention behaviour of the poly(methacrylates) provided the chance of separating different methacrylate homopolymers through normal-phase gradient elution. Since in this case separation is accomplished with respect to chemical composition and not molar mass, a second chromatographic method must be used for the determination of the MMD of the chromatographic fractions.

The general behaviour of a binary blend in different chromatographic modes is summarized in Fig. 6.40 [47]. In the size exclusion and the adsorption modes for both blend components, their retention behaviour is very similar, and the calibration curves log

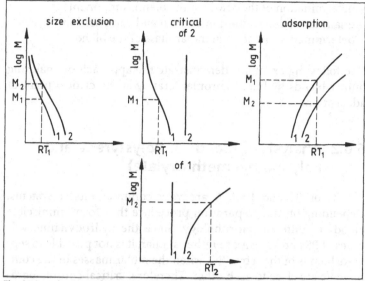

Fig. 6.40. Behaviour of the calibration curves log molar mass vs retention time of a binary polymer blend in the three chromatographic modes (Reprinted from Ref. [47] with kind permission from Elsevier Science Ltd., UK)

molar mass vs retention time suggest that in these cases normally one retention time corresponds to two molar masses (one molar mass on each calibration curve). Thus, for comparable molar masses of the components, an overlapping of the elution zones is obtained. Accordingly, a sufficient separation of the components using SEC or adsorption chromatography may be achieved only when their molar masses are quite different.

A completely different behaviour of the blend components is obtained when chromatographic conditions are used, corresponding to the critical mode of one of the components. In this case the elution zones are separated from each other over the entire molar mass range, and separation is achieved even for components of similar molar mass.

This type of separation holds much promise for the development of general separation schemes of polymer blends. With chromatographic conditions corresponding to the critical point of component 2, separation should be possible in any case. In addition, since component 1 is eluted in the SEC mode, its molar mass distribution can be determined. Changing the chromatographic system to the critical point of component 1, one can analyse component 2 with respect to its MMD.

For the analysis of binary blends the following protocol can be formulated:

– determination of the MMD of the blend components
– quantitative determination of the blend composition
– determination of the total molar mass of the blend.

The following examples demonstrate the approach of analysing binary blends by liquid chromatography at the critical point of adsorption.

6.6.2 Analysis of Blends of Polystyrene and Poly(methyl methacrylate)

Aim

Blends of PS and PMMA are very common model systems. Depending on the preparation procedure they form immiscible blends of different morphology. Since the hydrodynamic volumes of PS and PMMA are rather similar, it is not possible to separate blends of them by SEC, when the molar masses of the components are close to each other. Therefore, critical chromatography shall be used to separate PS-PMMA blends.

Materials

Calibration Standards. Narrow-disperse polystyrene (PS) and polymethyl methacrylate (PMMA) in the molar mass range of 1 000 to 300 000 g/mol.

Polymers. Blends of PS and PMMA of different compositions and molar masses. Such polymer blends can be easily prepared either by dissolving the components in a common solvent and evaporating the solvent in a film-forming procedure or by mixing the components in a Brabender Plasticorder.

Equipment

Chromatographic System. Modular HPLC system comprising a Waters Model 510 pump, a Rheodyne six-port injection valve and a Waters column oven.

Columns. Two column-set of LiChrospher Si-300 and Si-1000 (Merck), 10 µm average particle size and 300 Å and 1000 Å average pore size, with column sizes of 250x4.6 mm I.D., or a two column-set of Nucleosil RP-18 (Macherey-Nagel), 7 µm average particle size and 300 Å and 1000 Å average pore size, with column sizes of 250x4.6 mm I.D.

Mobile Phase. Mixtures of methylethylketone and cyclohexane or tetrahydrofuran and water. All solvents are HPLC grade.

Detectors. Waters 410 differential refractometer and Knauer fixed wavelength UV /vis detector at a wavelength of 280 nm.

Column Temperature. 25 °C.

Sample Concentration. 1–5 mg/mL.

Injection Volume. 50–100 μL.

The single components of a polymer blend can be analysed
using critical chromatographic conditions for the respective
blend components. Accordingly, for blends of PS and PMMA the
critical points of adsorption of PS and PMMA must be found.
Following the discussion on the elution behaviour as a function
of column polarity, see Fig. 6.25, different stationary phases
must be selected for establishing the critical points of PS and
PMMA, respectively. Since PMMA is the more polar component,
a polar (silica gel) column is chosen for establishing its critical
point. PS is then eluted in the SEC mode. For establishing the
critical point of PS, however, a reversed stationary phase must
be used. PMMA is eluted in the SEC mode under these conditions.

Preparatory
Investigations

The behaviour of PMMA of different molar masses on silica gel
in eluents comprising methylethylketone and cyclohexane is giv-
en in Fig. 6.27A and B, Section 6.5.2. For the two-column set Si-
300+Si-1000 the critical point corresponds to a mobile phase
composition of methylethylketone-cyclohexane 70:30% by volume.

For establishing the critical point of PS, a reversed-phase
Nucleosil RP-18 is used (see Fig. 6.41A). The figure indicates that
at concentrations of tetrahydrofuran >88% by volume, separation is predominantly driven by entropic effects and the SEC
mode is operating. In contrast, the LAC mode with predominantly enthalpic interactions is operating at concentrations of
tetrahydrofuran <87% by volume in the eluent. The critical point
of adsorption of PS is obtained at an eluent composition of
tetrahydrofuran-water 88.8:11.2% by volume. At this point, all PS
samples regardless of their molar mass elute at one retention
time. This, by definition, indicates that the PS polymer chain
behaves "chromatographically invisible", i.e., does not contribute
to retention. Accordingly, using these chromatographic conditions, blends of PMMA and PS can be analysed with respect to
the PMMA component.

Depending on the size of the macromolecules under investigation, similar to conventional SEC, the pore size of the packing
must be adjusted to the desired molar mass range. Thus, for
higher molar mass samples the investigations must be carried
out on column sets with larger pores, see Fig. 6.41B for a two-column set of RP-18 300 +1000 Å. For this column set, critical con-

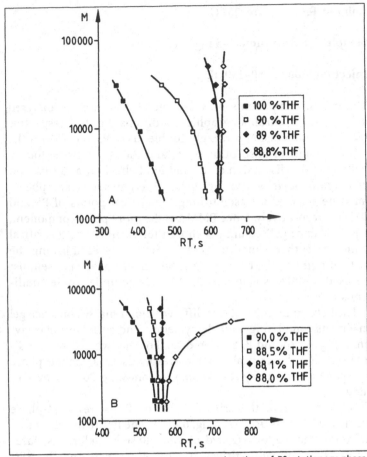

Fig. 6.41. Critical diagrams molar mass vs retention time of PS; stationary phase: Nucleosil RP-18 100 Å (A) or RP-18 300Å+1000 Å (B); mobile phase: tetrahydrofuran-water

ditions were found to be operating at an eluent composition of tetrahydrofuran-water 88.1:11.9% by volume.

Separations

In a first set of experiments, PS-PMMA blends are separated under chromatographic conditions, corresponding to the critical point of PS. By the use of Nucleosil RP-18 with an average pore size of 100 Å, the blends are completely separated into their components, see Fig. 6.42. Although the blends under investigation are composed of PS and PMMA of similar molar masses, a complete separation is obtained.

With a refractive index detector, a pronounced negative solvent peak is obtained between the elution peaks of the polymer com-

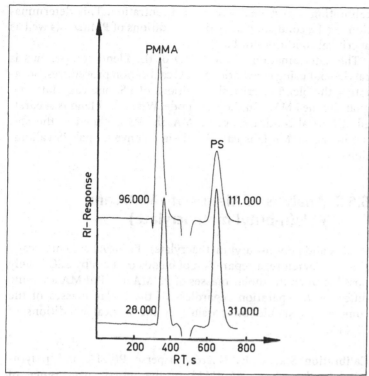

Fig. 6.42. Separation of binary blends of PS and PMMA at the critical point of PS; stationary phase: Nucleosil RP-18 100 Å: mobile phase: tetrahydrofuran-water 88.8:11.2% by volume

ponents. This solvent peak appears at $K_d=1$ due to preferential adsorption of tetrahydrofuran. Depending on the concentration and the polarity of the blend components, the solvent peak may partially or completely overlap one of the component peaks, thus interfering with a proper quantification of the elution peaks (see also Section 6.6.3).

The separation of the PS-PMMA blends can also be carried out under conditions corresponding to the critical point of PMMA. In this case, the columns are Nucleosil Si-300 + Si-1000 with a mobile phase of methylethylketone-cyclohexane 70:30% by volume. Since separation is carried out on a polar stationary phase, first the PS is eluted followed by the PMMA. The elution behaviour of PS corresponds to the SEC mode.

The quantitative analysis of the PS-PMMA blends is rather straightforward. The blend composition, i.e., the amounts of the components PS and PMMA, is determined via corresponding

Evaluation

calibration curves peak area vs concentration. This determination can be conducted at critical conditions of PMMA, as well as at critical conditions of PS.

The determination of the MMD of the blend components is carried out using conventional SEC calibration procedures. Separating the blend at critical conditions of PS, one can elute and quantify the PMMA in the SEC mode. When the blend is separated at critical conditions of PMMA, the PS is eluted in the SEC mode and its MMD is calculated via a conventional PS calibration curve.

6.6.3 Analysis of Blends of PMMA and Poly(n-butyl methacrylate)

Aim

PMMA and poly(n-butyl methacrylate) (PnBMA) are chemically similar. Therefore, a separation of blends of them by SEC is only possible when the molar masses of PMMA and PnBMA are quite different. A separation regardless of the molar masses of the components should be possible at the critical conditions of PMMA.

Materials

Calibration Standards. Narrow-disperse PMMA and poly(n-butyl methacrylate) (PnBMA) in the molar mass range of 1 000–300 000 g/mol.

Polymers. Blends of PMMA and PnBMA of different compositions and molar masses. Such polymer blends can be easily prepared either by dissolving the components in a common solvent and evaporating the solvent in a film-forming procedure or by mixing the components in a Brabender Plasticorder.

Equipment

Chromatographic System. Modular HPLC system comprising a Waters Model 510 pump, a Rheodyne six-port injection valve and a Waters column oven.

Columns. Critical Chromatography: Two column-set of LiChrospher Si-300 and Si-1000 (Merck), 10 µm average particle size and 300 Å and 1000 Å average pore size, with column sizes of 250x4.6 mm I.D., SEC: six column set of Ultrastyragel $10^6 + 2x10^5 + 2x10^4 + 10^3$ Å, 10 µm average particle size and columns sizes of 300x8 mm I.D.

Mobile Phase. Critical Chromatography: Mixtures of methylethylketone and cyclohexane, SEC: tetrahydrofuran. All solvents are HPLC grade.

Detectors. Waters 410 differential refractometer and a Viscotek 200 viscometer detector.

Column Temperature. 25 °C.

Sample Concentration. 1–5 mg/mL.

Injection Volume. 50–100 μL for CC and SEC.

The critical point of PMMA on silica gel Nucleosil Si-300 + Si-1000 corresponds to a mobile phase composition of methylethylketone 70:30% by volume (see Section 6.5.2).   Preparatory Investigations

The high-resolution SEC chromatogram of a technical blend of PMMA and PnBMA is given in Fig. 6.43. A completely uniform unimodal elution peak is obtained, not indicating any presence of two or more components, although the molar masses of the components are significantly different (PMMA ~ 50 000 g/mol, PnBMA ~ 90 000 g/mol).   Separations

If now similar blends are investigated by chromatography at the critical point of adsorption of PMMA, a complete separation into components is obtained. Figure 6.44A shows the separation of blends of different compositions using a refractive index detector. A negative solvent peak is obtained owing to preferential adsorption. Unfortunately, there is very strong overlapping of the solvent peak and the elution peak of the PMMA, making it impossible to quantify the PMMA peak.

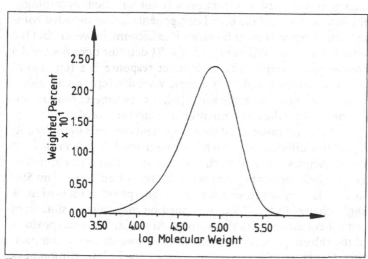

Fig. 6.43. SEC chromatogram of a blend of PMMA and PnBMA; stationary phase: Ultrastyragel; mobile phase: THF

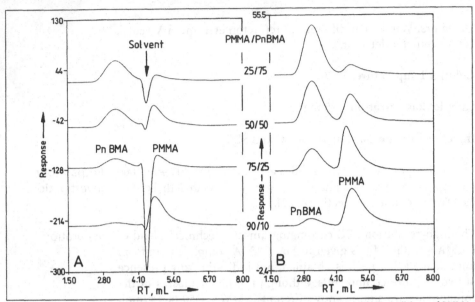

Fig. 6.44. Chromatograms of PMMA-PnBMA blends of different composition at critical conditions of PMMA; stationary phase: Nucleosil Si-300 + Si-1000; mobile phase: methyl ethyl ketone-cyclohexane 70:30% by volume; detector: differential refractometer (A) or viscometer (B)

Much more informative chromatograms are obtained, when instead of a RI detector an on-line viscometer is used, see Fig. 6.44B. Since a slight change in the composition of the mobile phase due to preferential adsorption does not contribute to a change in viscosity, a solvent peak is not obtained. Accordingly, the elution peaks of the blend components can be detected without interference. It must be taken into account, however, that the viscosity detector different from the RI detector does not yield a concentration signal. The viscometer response is a function of the concentration and the intrinsic viscosity ($c[\eta]$), and unless both components have the same $[\eta]$, the viscometer output cannot be used for the determination of concentrations.

For the determination of the component concentrations by a RI detector a different approach may be helpful. It is known from the principles of critical chromatography that the chromatographic behaviour of a polymer in the transition range from SEC to LAC is very sensitive towards mobile phase composition. A slight change in the eluent composition may cause a shift from critical conditions into the adsorption mode. Since the position of the solvent peak is rather insensitive towards eluent composition, this fact could be used to separate the PMMA elution peak from the solvent peak.

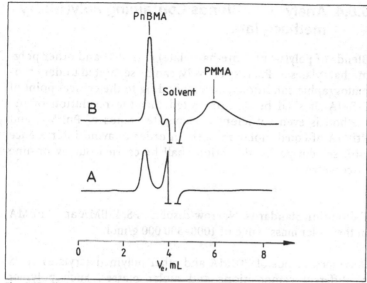

Fig. 6.45. Chromatograms of a PMMA-PnBMA blend at different compositions of the mobile phase; stationary phase: Nucleosil Si-300 + Si-1000; mobile phase: methyl ethyl ketone-cyclohexane 70:30% (A) or 68:32% by volume (B)

Figure 6.45 demonstrates the changes in the chromatogram when changing the mobile phase composition from methyl ethyl ketone-cyclohexane 70:30 to 68:32% by volume, corresponding to a slight adsorption mode. In this case the PMMA elution peak is well separated from the solvent peak and quantification can be carried out. The PnBMA elutes in the SEC mode and the solvent peak appears between the elution peaks of the components.

The quantitative analysis of the PMMA-PnBMA blends is done   Evaluation
similar as described in Section 6.6.2. The blend composition, i.e., the amounts of the components PnBMA and PMMA, is determined via corresponding calibration curves peak area vs concentration.

The determination of the MMD of the PnBMA is carried out using conventional SEC calibration procedures. Separating the blend at critical conditions of PMMA, the PnBMA is eluted in the SEC mode and quantified accordingly. For the determination of the MMD of PMMA a chromatographic system must be used, where at the critical point of PnBMA the PMMA is eluted in the SEC mode.

6.6.4 Analysis of Blends Containing Poly(t-butyl methacrylate)

Aim

Blends of poly(t-butyl methacrylate) (PtBMA) and other poly-methacrylates or PS, respectively, can be separated under chromatographic conditions, corresponding to the critical point of PtBMA. It shall be demonstrated that the resolution of the method is even sufficient to separate blends of PnBMA and PtBMA of equal molar masses. In order to avoid interference with solvent peaks, detection shall be carried out by on-line viscometry.

Materials

Calibration Standards. Narrow-disperse PS, PtBMA and PnBMA in the molar mass range of 1000–300 000 g/mol.

Polymers. Blends of PtBMA and other polymethacrylates or PS of different compositions and molar masses. Such polymer blends can be easily prepared either by dissolving the components in a common solvent and evaporating the solvent in a film-forming procedure or by mixing the components in a Brabender Plasticorder.

Equipment

Chromatographic System. Modular HPLC system comprising a Waters Model 510 pump, a Rheodyne six-port injection valve and a Waters column oven.

Columns. Two column-set of LiChrospher Si-300 and Si-1000 (Merck), 10 µm average particle size and 300 Å and 1000 Å average pore size, with column sizes of 250x4.6 mm I.D.

Mobile Phase. Mixtures of methyl ethyl ketone and cyclohexane. All solvents are HPLC grade.

Detectors. Viscotek 200 viscometer detector.

Column Temperature. 25 °C.

Sample Concentration. 1–5 mg/mL.

Injection Volume. 50–100 µL.

Preparatory Investigations

Similar to the critical point of PMMA, the critical point of PtBMA can be established on silica gel Nucleosil Si-300 + Si-1000. The critical point of PtBMA corresponds to a mobile phase composition of methylethylketone 18.8:81.2% by volume (see Fig. 6.46).

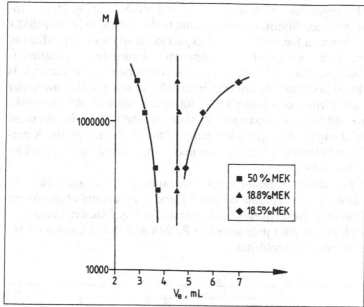

Fig. 6.46. Critical diagram molar mass vs elution volume of poly(t-butyl methacrylate); stationary phase: Nucleosil Si-300 + Si-1000; mobile phase: methyl ethyl ketone-cyclohexane; detector: viscometer

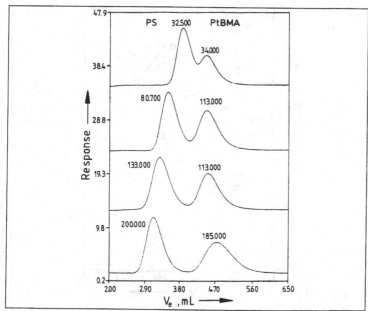

Fig. 6.47. Separation of blends of PS and PtBMA at the critical point of PtBMA; stationary phase: Nucleosil Si-300 + Si-1000; mobile phase: methyl ethyl ketone-cyclohexane 18.8:81.2% by volume; detector: viscometer

Separations

The separation of blends of PS and PtBMA using chromatographic conditions, corresponding to the critical point of PtBMA is shown in Fig. 6.47. As was expected, in all cases, regardless of the molar masses of the components, a complete separation is obtained. Since PS is the less polar component in the blend, it is eluted first from the column in the SEC mode. For the low molar mass blend, comprising PS 32 500 and PtBMA 34 000, the peaks are not baseline-separated. A better resolution can be obtained by changing the separation range of the stationary phase. A useful combination for this molar mass range would be Nucleosil Si-100 + Si-300.

The ability of liquid chromatography at the critical point of adsorption to separate polymer blend components of minimum structural differences is demonstrated in Fig. 6.48. Even blends of such very similar polymers like PtBMA and PnBMA can be separated without problems.

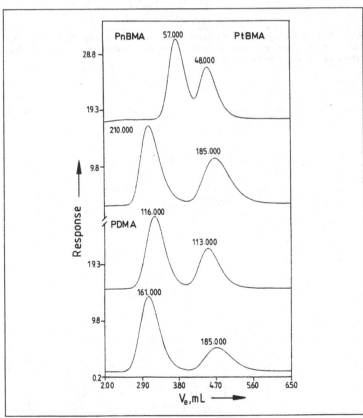

Fig. 6.48. Separation of blends PtBMA and PDMA or PnBMA at the critical point of PtBMA; stationary phase: Nucleosil Si-300 + Si-1000: mobile phase: methyl ethyl ketone-cyclohexane 18.8:81.2% by volume

The quantitative composition of the blends cannot be directly obtained from the viscometer signal. The viscometer output is a function of the concentration and the intrinsic viscosity, and therefore, the concentration can only be determined, when the intrinsic viscosity of the components is known. However, for obtaining the component elution peaks without interference of solvent peaks, the viscometer is a useful tool.

Evaluation

References

1. PASCH H, MUCH H, SCHULZ G, GORSHKOV A V (1992) LC-GC Int 5:38
2. MUENKER AH, HUDSON BE (1969) J Macromol Sci A3: 1465
3. EVREINOV VV (1970) Vysokomol Soedin A12: 829
4. ENTELIS SG, EVREINOV VV, GORSHKOV AV (1986) Adv Polym Sci 76: 129
5. GORSHKOV AV, MUCH H, BECKER H, PASCH H, EVREINOV VV, ENTELIS SG (1990) J Chromatogr 523: 91
6. GORSHKOV AV (1985) Vysokomol Soedin B27: 181
7. PASCH H, ZAMMERT I (1994) J Liquid Chromatogr 17: 3091
8. PASCH H, ZAMMERT I, JUST U (1995) J Polym Anal Char, 1: 329
9. KARAS M, HILLENKAMP F (1988) Anal Chem 60: 2299
10. BEAVIS RC (1989) Rapid Commun Mass Spectrom 3: 233
11. PASCH H, UNVERICHT R (1993) Angew Makromol Chem 212: 191
12. KRÜGER R-P, MUCH H, SCHULZ G (1994) J Liquid Chromatogr 17: 3069
13. VAKHTINA IA, OKUNIEVA AG, TECHRITZ R, TARAKANOV O G (1976) Vysokomol Soedin A18: 471
14. FILATOVA NN, ROVINA DY, EVREINOV VV, ENTELIS SG (1978) Vysokomol. Soedin A20: 2367
15. FILATOVA NN, GORSHKOV AV, EVREINOV VV, ENTELIS SG (1988) Vysomol Soedin A30: 953
16. REMPP P, FRANTA E (1984) Adv Polym Sci 58: 3
17. HEITZ W (1986) Angew Makromol Chem 145/146: 37
18. KRÜGER H, MUCH H, SCHULZ G, WEHRSTEDT C (1990) Makromol Chem 191: 907
19. KRÜGER H, PASCH H, MUCH H, GANCHEVA V, VELICHKOVA R (1992) Makromol Chem 193: 1975
20. PASCH H, KRÜGER H, MUCH H, JUST U (1992) J Chromatogr 589: 295
21. GLÖCKNER G (1991) Gradient HPLC of Copolymers and Chromatographic Cross-Fractionation. Springer, Berlin Heidelberg New York
22. TERAMACHI S, HASEGAWA A, SHIMA Y, AKATSUKA M, NAKAJIMA M (1979) Macromolecules 12: 992
23. GLÖCKNER G, VAN DEN BERG JHM (1986) J Chromatogr 352: 511
24. AUGENSTEIN M, STICKLER M (1990) Makromol Chem 191: 415
25. GANKINA E, BELENKII B, MALAKHOVA I, MELENEVSKAYA E, ZGONNIK V (1991) J Planar Chromatogr 4: 199

26. ZIMINA T M, KEVER J J, MELENEVSKAYA E Y, FELL A F (1992) J Chromatogr 593: 233
27. PASCH H, BRINKMANN C, GALLOT Y (1993) Polymer 34: 4100
28. PASCH H, GALLOT Y, TRATHNIGG B (1993) Polymer 34: 4986
29. TRATHNIGG B (1991) In: Lemstra PJ, Kleintjens LA (eds) Integr Fundam Polym Sci Technol Proc Int Meet Polym Sci Technol, vol 5. Elsevier Applied Science, London
30. TRATHNIGG B, YAN X (1992) Chromatographia 33: 467
31. PASCH H, AUGENSTEIN M, TRATHNIGG B (1994) Macromol Chem Phys 195: 743
32. PASCH H, AUGENSTEIN M (1993) Makromol Chem 194: 2533
33. AUGENSTEIN M, MÜLLER MA (1990) Makromol Chem 191: 2151
34. GORBUNOV AA, SKVORTSOV AM (1988) Vysokomol Soedin A30: 895
35. PASCH H, RODE K (1995) J Chromatogr 699: 21
36. PASCH H, KRÜGER H, MUCH H, JUST U (1992) Polymer 33: 3889
37. GREGORIOU VS, NODA I, DOWREY AI, MARCOTT C, CHAO JL, PALMER PA (1993) J Polym Sci Polym Phys 31: 1769
38. NISHIOKA T, TERAMAE N (1993) J Appl Polym Sci Polym Symp 52: 251
39. GHEBREMESKEL Y, FIELDS J, GARTON A (1994) J Polym Sci Polym Phys 32: 383
40. SCHMIDT P, DYBAL J, STRAKA J, SCHNEIDER B (1993) Makromol Chem 194: 1757
41. KELTS LW, LANDRY CJ, TEENGARDEN DM (1993) Macromolecules 26: 2941
42. HUO PP, CEBE P (1993) Macromolecules 26: 5561
43. CANTOW HJ, PROBST J, STOJANOV C (1968) Kautschuk Gummi 21: 609
44. SCHRÖDER E, FRANZ J, HAGEN E (1976) Ausgewählte Methoden der Plastanalytik, Akademie-Verlag, Berlin
45. TRATHNIGG B, YAN X (1992) Chromatographia 33: 467
46. MOUREY TH (1986) J Chromatogr 357: 101
47. PASCH H (1993) Polymer 34: 4095
48. KILZ P, GORES F (1993) In: Provder T (ed) Chromatography of Polymers. ACS Symp. Ser. 521, ACS, Washington, Chapter 10

7 Two-Dimensional Liquid Chromatography

7.1 Peculiarities

Complex polymers are distributed in more than one direction. Copolymers are characterized by the molar mass distribution and the chemical heterogeneity, whereas functional homopolymers are distributed in molar mass and functionality. Hence, the experimental evaluation of the different distribution functions requires separation in more than one direction.

The classical approach is based upon the dependence of copolymer solubility on composition and chain length. A solvent/non-solvent combination fractionating solely by molar mass would be appropriate for the evaluation of MMD, another one separating with respect to chemical composition would be suited for determining CCD or FTD. A clear presentation of the situation for random copolymers was given by Rosenthal and White (see Fig. 7.1) [1].

The precipitation fractionation of cellulose acetate yielded fractions which varied both in acetyl content and intrinsic viscosity. As expected, fractionation was influenced by MMD and CCD (see Figs. 7.1d and 7.1e), and as a result fractions were obtained which were similarly distributed in MMD and CCD. Even high resolution fractionation would not improve the result (see Fig. 7.1f). Narrower fractions can be obtained by cross-fractionation separating in two different directions (see Fig. 7.1 g). However, even in this case it is nearly impossible to obtain perfectly homogeneous fractions.

By the use of different modes of liquid chromatography it is possible to separate polymers selectively with respect to hydrodynamic volume (molar mass), chemical composition or functionality. Using these techniques and combining them with each other or with a selective detector, two-dimensional information on different aspects of molecular heterogeneity can be obtained. If, for example, two different chromatographic techniques are combined in a "cross-fractionation" mode, information on CCD and MMD can be obtained. Literally, the term "chromatographic cross-fractionation" refers to any combination of chromatographic methods capable of evaluating the distribution in size

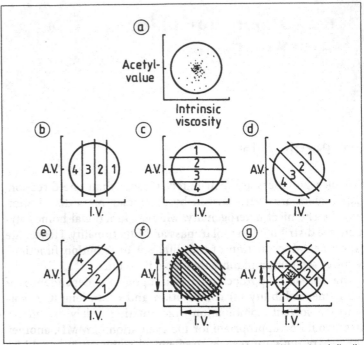

Fig. 7.1. Fractionation and cross-fractionation schemes. **a** Two-dimensional distribution acetyl value (chemical composition) vs intrinsic viscosity (molar mass) of a cellulose acetate sample, **b** fractionation according to molar mass, **c** fractionation according to chemical composition, **d, e** fractionation according both to molar mass and chemical composition depending on the solvent/non-solvent combination, **f** high-resolution fractionation, **g** cross- fractionation using two different solvent/non-solvent combinations (Reprinted from Ref. [1] with permission)

and composition of copolymers. An excellent overview on different techniques and applications involving the combination of SEC and gradient HPLC was published by Glöckner [2].

The molecular size of a copolymer molecule in solution is a function of its chain length, chemical composition, solvent and temperature. Thus, molecules of the same chain length but different composition have different hydrodynamic volumes. Since SEC separates according to hydrodynamic volume, SEC in different eluents can separate a copolymer in two diverging directions. This principle of "orthogonal chromatography" was suggested by Balke and Patel [3]. The authors coupled two SEC instruments together so that the eluent from the first one flowed through the injection valve of the second one. At any desired retention time the flow through SEC 1 could be stopped and an injection made into SEC 2. The first instrument was operated with THF as the eluent and polystyrene gel as the packing, whereas for SEC 2 polyether bonded-phase columns and THF-heptane were used.

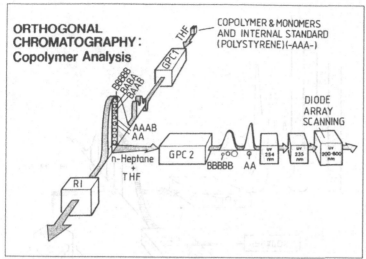

Fig. 7.2. Schematic presentation of an orthogonal chromatographic system showing size fractionation of a linear copolymer by SEC 1 and the variety of molecules of the same molecular size within a chromatogram slice, A-styrene and B-butyl methacrylate units

The schematic presentation of this system is given in Fig. 7.2. Both instruments utilized GPC columns. However, whereas the first GPC was operating so as to achieve conventional molecular size separation, the second GPC was used to fractionate by composition, utilizing a mixed solvent to encourage adsorption and partition effects in addition to size exclusion.

The authors reported the investigation of random copolymers of styrene and n-butyl methacrylate, containing the parent homopolymers PS and PnBMA. While in SEC 1 fractions of different molecular size were obtained, a separation with respect to chemical composition into fractions of PnBMA, P(St/nBMA) and PS could be achieved in SEC 2, the elution order being PnBMA <P(St/nBMA)<PS. The explanation of this behaviour is a synergistic effect of different separation mechanisms including size exclusion, adsorption and partition.

Much work on chromatographic cross-fractionation was carried out with respect to combination of SEC and gradient HPLC. In most cases SEC was used as the first separation step, followed by HPLC, as illustrated in Fig. 7.3. In a number of early papers the cross-fractionation of model mixtures was discussed. Investigations of this kind demonstrated the efficiency of gradient HPLC for separation by chemical composition. Mixtures of random copolymers of styrene and acrylonitrile were separated by Glöckner et al. [5]. In the first dimension a SEC separation was

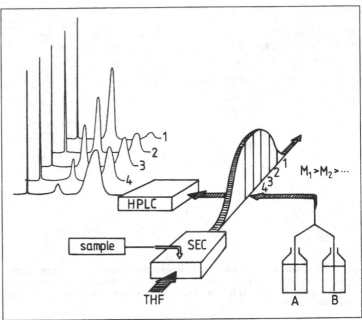

Fig. 7.3. Chromatographic cross-fractionation by SEC prefractionation and gradient HPLC of eluate fractions without additional treatment; first peak from the left indicates a solvent peak (Reprinted from Ref. [2] with permission of Springer-Verlag)

carried out using THF as the eluent and polystyrene gel as the packing. In total, about 10 fractions were collected and subjected to the second dimension, which was gradient HPLC on a CN bonded-phase using isooctane/THF as the mobile phase. Model mixtures of random copolymers of styrene and 2-methoxyethyl methacrylate were separated in a similar way, the mobile phase of the HPLC mode being isooctane/methanol in this case [6]. This procedure was applied to real-world copolymers as well [5]. Graft copolymers of methyl methacrylate onto EPDM rubber were analysed by Augenstein and Stickler [7]; whereas, Mori reported on the fractionation of block copolymers of styrene and vinyl acetate [8].

A more feasible way of analysing copolymers is the prefractionation through HPLC in the first dimension and subsequent analysis of the fractions by SEC [9, 10]. HPLC was found to be rather insensitive towards molar mass effects and yielded very uniform fractions with respect to chemical composition.

The major disadvantage of all early investigations on chromatographic cross-fractionation was related to the fact that both separation modes were combined to each other either off-line or in a stop-flow mode. Regardless of the separation order SEC vs

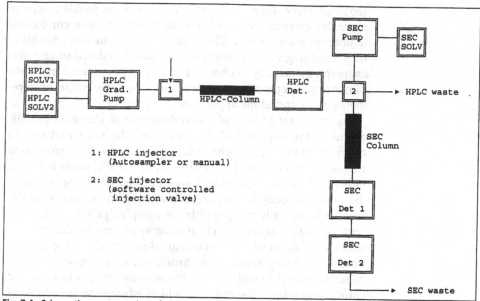

Fig. 7.4. Schematic representation of an automated two-dimensional chromatographic system

HPLC or HPLC vs SEC, in the first separation step fractions were collected, isolated, and then subjected to the second separation step. This procedure, of course, is very time-consuming and the reliability of the results at least to a certain extent depends on the skills of the operator.

A fully automated two-dimensional chromatographic system was developed recently by Kilz et al. [11–13]. It consists of two chromatographs, one which separates by chemical composition or functionality and a SEC instrument for subsequent separation by size. Via a storage loop system, fractions from the first separation step are automatically transfered into the second separation system. The operation of the column switching device is automatically driven by the software, which at the same time organizes the data collection from the detector. The design of this system is presented schematically in Fig. 7.4.

7.2 Equipment and Materials

The analysis of complex polymers by two-dimensional chromatography can be carried out in off-line and on-line modes. Regardless of the sequence of the different separation modes, two different chromatographic systems comprising an eluent reservoir, a HPLC pump, an injector, and a column combination are required. Depending on the specific separation problem, one or more detectors may be used in both or only in the second chromatographic dimension. The requirements to the components of the chromatographic systems are discussed in detail in previous chapters. Principally, a chromatographic system used in a two-dimensional separation scheme must meet the same requirements as a stand-alone apparatus.

An important feature of a two-dimensional chromatographic system is the sequence of the separations to be carried out. In many cases interaction chromatography as the first separation step is the best choice. The separation in the SEC mode is always influenced by the chain length and the chemical composition of the macromolecules; a separation solely with respect to molar mass is principally not possible for complex polymers. In contrast to SEC, interaction chromatography can be carried out under conditions where molar mass does not affect the separation. Thus, using adsorption chromatography, copolymers can be separated solely with respect to chemical composition. Liquid chromatography at the critical point of adsorption is capable of separating exclusively by functional groups. It is therefore, more feasible to separate with respect to CCD or FTD in the first separation step, followed by SEC in the second dimension.

In addition, from an experimental point of view, high flexibility is required for the first chromatographic dimension. This is better achieved by running the interaction chromatography mode in the first dimension due to the fact that (1) more parameters (mobile phase, stationary phase, temperature) can be used to adjust the separation according to the chemical nature of the sample, (2) better fine-tuning in HPLC gives more homogeneous fractions, and (3) sample load on HPLC columns can be much higher as compared to SEC columns.

The experimental treatment of two-dimensional chromatography depends on the availability of chromatographic components. The simplest method is the off-line analysis of fractions obtained by separating a sample with respect to chemical composition or functionality (see Fig. 7.5).

The fractions are collected and then separately analysed by SEC. The advantage of this procedure is that it can be carried out using basic chromatographic equipment. The fractions are processed separately, i.e., the eluent of the first separation step may be removed, and the fractions may be modified chemically, if necessary. However, this technique is time-consuming and relies on the skills of the operator.

The simplest way to an automated system is given in Fig. 7.6. In this case, the two chromatographic systems are connected to each other via a storage loop and an additional injection valve.

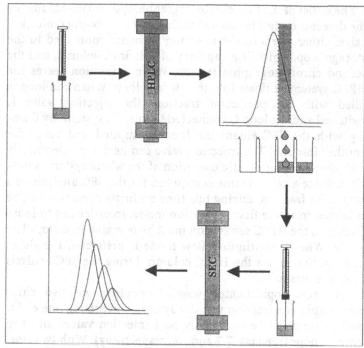

Fig. 7.5. Schematic representation of two-dimensional chromatography in the off-line mode

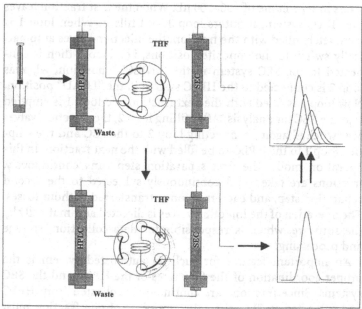

Fig. 7.6. Schematic representation of two-dimensional chromatography in the on-line mode using one storage loop

The outlet of the first chromatographic system (the column or the detector outlet) is coupled to the inlet of a six-port injection valve. Other inlets of the injection valve are connected to the storage loop, which is a capillary of a defined volume, and the second chromatographic system. When a fraction leaves the HPLC system, it flows into the storage loop. When the loop is filled with the particular fraction, the injection valve is switched and the loop is connected to the SEC system. By flushing with the SEC eluent the loop is emptied and ready for another fraction. The injection valve can be driven electrically and, therefore, automatic operation of the whole system is possible. Since a certain time is required for the SEC analysis of a particular fraction, during this time no further fraction can be collected from the first separation mode. In order not to loose fractions, the HPLC separation must be operated in a stop-flow mode. When a continuous-flow mode is preferred, the eluate fraction that leaves the HPLC column during the SEC analysis goes into waste.

The most sophisticated way of coupling the two chromatographs is a fraction transfer system, comprising one eight-port injection valve or two six-port injection valves and two storage loops (see Figs. 7.7 and 7.8, respectively). With two storage loops, it is possible to collect fractions continuously without losses. When starting the separation, loop 1 is in the "LOAD" position, whereas loop 2 is in the "INJECT" position. Each of the loops has a volume of 100–200 μL. When the first fraction leaves the HPLC system, it enters loop 1 and fills it. When loop 1 is completely filled with the fraction, the injection valves automatically switch to the opposite positions, i.e., loop 1 then is connected to the SEC system in the "INJECT" position, whereas loop 2 is connected to the HPLC system in the "LOAD" position. Now loop 2 is filled with the next fraction and loop 1 is emptied into the SEC for analysis. After filling loop 2, the injection valves are switched again, connecting loop 2 to the SEC and the emptied loop 1 to the HPLC to be filled with the next fraction. In this operation mode, the first separation step runs continuously, fractions are taken and continuously subjected to the second separation step, and each fraction is transferred without losses. The operation of the injection valves is directed automatically by the software, which is responsible for data collection, storage and processing.

An important feature for such an automated system is the proper coordination of the flow rates of the HPLC and the SEC systems. Since fractions are continuously collected from HPLC and subjected to SEC, the collection time of one fraction must exactly equal the analysis time in the SEC mode. Depending on

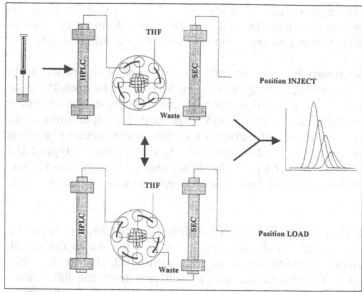

Fig. 7.7. Schematic representation of two-dimensional chromatography in the online mode using a eight-port injection valve and two storage loops

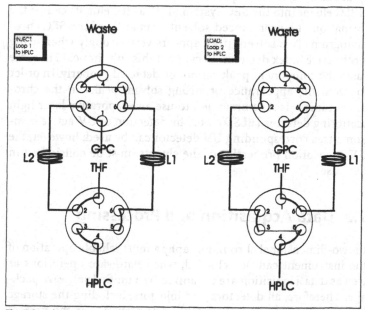

Fig. 7.8. Schematic representation of two-dimensional chromatography in the online mode using two six-port injection valves and two storage loops

the number and size of the SEC columns, about 7–15 min are required for one SEC analysis. The flow rate in HPLC must be such, that one storage loop is filled exactly within this time.

Stationary Phases. Depending on the type of separation in the HPLC mode, packings of varying polarity can be used. If separations are conducted in the adsorption mode, stationary phases described in Chapter 3 and 5 are used. For separations in the critical mode of adsorption, the corresponding packing described in Chapter 6 are used. In the SEC mode, typical SEC columns are used (see Chapters 3 and 4). Usually, two to three columns of cross-linked polystyrene gel give sufficient resolution.

Mobile Phases. For the HPLC mode, refer to Chapters 3, 5 and 6. For the SEC mode, in most cases THF is used as the mobile phase. Attention has to be paid to the fact, that in the HPLC mode the eluents are usually different from the SEC eluent. Therefore, one has to make sure that there are no miscibility problems between the eluents of the HPLC and SEC modes.

Detectors. Each fraction from the HPLC separation represents a very dilute solution. When injecting this fraction into the SEC system, further dilution takes place. Accordingly, high sensitivity of the detector is required. In addition, the injection of the HPLC eluent into the SEC system (THF as the eluent) causes the formation of a pronounced solvent peak at V_0 of the SEC chromatogram. This solvent peak appears very strongly when using a refractive index detector. Because of this solvent peak, in some cases the component peak cannot be detected properly. In order to avoid the appearance of strong solvent peaks in the chromatograms it is recommended to use an evaporative laser light scattering detector (ELSD). For the detection of UV active components, a corresponding UV detector can be used; however, the UV absorption properties of the eluents must be considered in this case.

7.3 Data Acquisition and Processing

In two-dimensional chromatography a most reliable operation of the instrument can be achieved, when hardware operations as well as data acquisition are organized by a unified software package. Therefore, all detectors and injectors, including the storage loops, are connected to a suitable data station. For the present investigations the 2D-CHROM hardware and software of Polymer Standards Service, Mainz, Germany, is used. It allows one to operate the switching valves of the storage loops by suitable time events, to conduct data acquisition and all calculations.

In addition to qualitative information on the molecular heterogeneity of a complex copolymer, quantitative data on the different distribution functions must be obtained. Since the first step preferably is HPLC, a calibration with respect to chemical composition or functionality has to be conducted. The quantitative determination of FTD is rather straightforward, because a separation into homogeneous functionality fractions is obtained by liquid chromatography at the critical point of adsorption. These fractions must be identified and can then be quantified via the corresponding peak areas taking into account the respective response factors. For a more detailed description of this procedure, refer to Section 6.4.

Calibration and Quantification

The quantitative determination of CCD depends on the separation procedure. If a separation into single oligomers is obtained, these may be quantified via the corresponding peak areas. The determination of the chemical composition of diblock and triblock copolymers can be carried out according to the procedure described in Section 6.5. The separation of random copolymers with respect to CCD is possible by gradient HPLC. In this case, the quantitative evaluation of the copolymer chromatograms requires knowledge of the influence of polymer composition on elution time, the influence of molar mass on elution time, and the influence of composition on detector signal intensity. This information can be obtained by calibration with a series of samples graded in composition. For more details, see [2].

The calibration of the second separation step, which is in most cases SEC, can be carried out similar to ordinary SEC experiments; see Chapter 4 for more details. However, a number of important features of two-dimensional chromatography must be considered: every calibration procedure relates a certain elution volume to a certain molar mass. This calibration is valid as long as the chromatographic procedure and apparatus are not changed. If now, instead of an ordinary injection of a polymer sample into the SEC eluent, an injection of a polymer sample in a mixed solvent via an automated switching valve is made, the hydrodynamic volume and the elution volume of the sample will change. Therefore, in order to obtain reproducible and reliable results, the SEC calibration standard must pass the first separation step (HPLC) and enter the SEC system via automated injection. The calibration of the SEC system has to be carried out in the following way:

1. The calibration standard is injected into the first chromatographic system and the retention volume V_1 is determined only for this system.
2. The calibration standard is injected into the two-dimensional system and after V_1 the automated injection valve is switched to inject the calibration standard into the SEC system. The injection time is taken as the start point ($V_e=0$) and the elution volume V_2 for the SEC system is determined.
3. V_2 is plotted against the molar mass of the calibration standard to yield the calibration curve of the SEC system.

The visualization of the results of a two-dimensional separation is possible in different ways. A discontinuous presentation is obtained, when the SEC chromatograms of all fractions are plotted along the elution volume axis of the first separation step (see Fig. 7.9).

A much clearer presentation of the separation is obtained when a continuous plot is used (see Fig. 7.10). In this case the separation in the first and second modes is plotted along the Y- and X-axis, respectively. The concentration profile is presented by a colour code in the Z-axis.

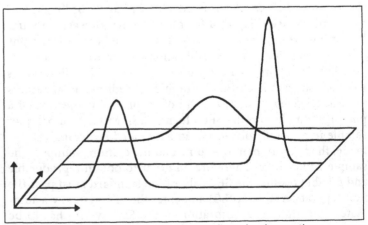

Fig. 7.9. Discontinuous representation of a two-dimensional separation

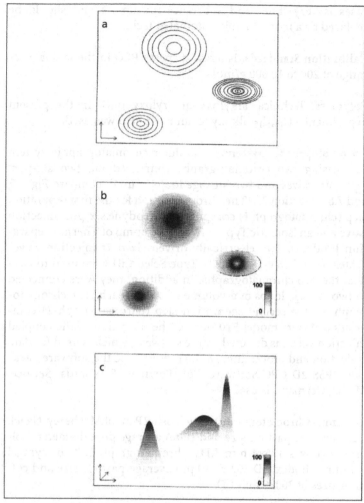

Fig. 7.10. Continuous representation of a two-dimensional separation: a intensity line diagram, b contour plot, c surface plot

7.4 Separation of a Technical Poly(ethylene oxide)

As outlined in Section 6.4.2, PEOs are heterogeneous with respect to functionality and molar mass. The functionality type distribution can be determined by liquid chromatography at the critical point of adsorption where the samples are separated into functionally uniform fractions. These fractions can then be analysed with respect to molar mass. This procedure in some

Aim

cases is very time- and substance-consuming and should be replaced by a more feasible automated technique.

<table>
<tr><td>Materials</td><td>

Calibration Standards. Narrow-disperse PEG in the molar mass range of 200 to 10 000 g/mol.

Polymers. Technical alkyloxy or aryloxy PEO. In the present experiment a C_{13},C_{15}-alkoxy terminated PEO was used.

</td></tr>
</table>

Materials

Calibration Standards. Narrow-disperse PEG in the molar mass range of 200 to 10 000 g/mol.

Polymers. Technical alkyloxy or aryloxy PEO. In the present experiment a C_{13},C_{15}-alkoxy terminated PEO was used.

Equipment

Chromatographic System. A modular chromatographic system comprising two chromatographs connected via two six-port injection valves and two storage loops is used, compare Fig. 7.7 and 7.8 in Section 7.2. The chromatograph for the first separation step (chromatograph 1) comprises a Rheodyne six-port injection valve and an isocratic Type P 100 HPLC pump of Thermo Separation Products. Two electrically driven six-port injection valves (Latek Type HMV-P and ECP Type Selec-Cil) were used to connect the two chromatographs. In addition, they were connected to two storage loops of a volume of 200 µL each. The chromatograph for the second separation step (chromatograph 2) comprises a Waters model 510 pump. The operation of the coupled injection valves is directed by the software, which is used for data collection and processing. In the present case the software package "PSS-2D-GPC-Software" of Polymer Standards Service, Mainz, Germany, is used.

Columns. Chromatograph 1: Nucleosil RP-18 of Macherey-Nagel, 5 µm average particle size and 100 Å average pore diameter. Column size was 125x4 mm I.D.. Chromatograph 2: Two styragel columns Shodex AD 802-S, 10 µm average particle size and column sizes of 300x8 mm I.D.

Mobile Phase. Chromatograph 1: Methanol-water, Chromatograph 2: THF, all solvents are HPLC grade

Detectors. Waters differential refractometer R 401 in chromatograph 1 and an evaporative light scattering detector (ELSD) model 950/40 of Zinsser Analytics in chromatograph 2.

Column Temperature. 25 °C.

Sample Concentration. 20–40 mg/mL. All samples are dissolved in the mobile phase of chromatograph 1.

Injection Volume. 50 µL.

In order to separate a functional homopolymer with respect to   Preparatory
the functional end groups, the separation must be conducted at   Investigations
conditions, corresponding to the critical point of adsorption of
the polymer chain. For PEO the critical point is obtained using a
reversed-stationary phase RP-18 and a mobile phase of acetonitrile-water 46:54% by volume (see Section 6.4.2). For PEOs with
long fatty alcohol endgroups, however, very long retention times
may be obtained. In order to increase desorption from the stationary phase and, therefore, reduce retention times, a mobile
phase of methanol-water was used. For this mobile phase, the
critical point of adsorption corresponds to a composition of
methanol-water 86:14% by volume.

In a first experiment the C_{13},C_{15}-alkoxy terminated PEO is sepa-   Separations
rated using chromatograph 1 only. Three well-separated peaks
appear in the chromatogram, indicating that separation takes
place with respect to the functionality fractions. Since separation
is very good, the mobile phase composition is changed from
methanol-water 86:14 to 88:12% by volume to decrease retention
times. For two-dimensional separations it is highly important to
elute the fractions as close as possible one after another, otherwise, a large number of fractions taken from chromatograph 1
and being transferred to chromatograph 2 would comprise only
of mobile phase. The separation of the sample in chromatograph 1
is presented in Fig. 7.11. The peaks of the functionality fractions

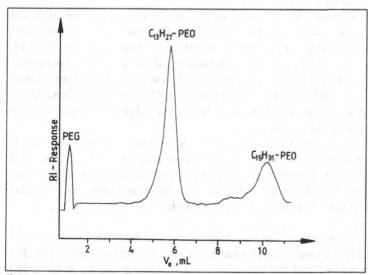

Fig. 7.11. Chromatogram of a C_{13},C_{15}-PEO sample separated in chromatograph 1;
stationary phase: RP-18, 125x4 mm I.D.; mobile phase: methanol-water 88: 12% by
volume

are well separated and the identification is carried out by mass spectrometry. In agreement with the expected elution order, the first peak is a PEG fraction; whereas, the second and third peaks contain the C_{13}-PEO and C_{15}-PEO fractions, respectively.

Once the sample is sufficiently well separated in the first separation step, a two-dimensional experiment can be carried out. In a two-dimensional experiment fractions are permanently taken from chromatograph 1, stored in the loops 1 and 2 and then transferred to chromatograph 2. To do this continuously, the storage loops must be filled in exactly the time which is needed to run one separation in chromatograph 2. In this manner, a fraction is collected from chromatograph 1 into loop 1, transferred to chromatograph 2, analysed, and the resulting chromatogram is stored. During this analysis time, loop 2 is filled with the next fraction and transferred to chromatograph 2 for analysis after analysing fraction 1.

Since for the present configuration, one analysis in chromatograph 2 is equal to an elution volume of about 10 mL, which equals to a retention time of 5 min at a flow rate of 2 mL/min, a loop of 200 µL must be filled with the next fraction within this time. For chromatograph 1, a flow rate of 40 µL/min must be used.

As a result of the two-dimensional separation, a number of chromatograms from chromatograph 2 are obtained, each of them characterizing a fraction of 200 µL from chromatograph 1. The chromatograms can be presented in a discontinuous plot as is shown in Fig. 7.12.

A much clearer presentation of the results can be obtained when a continuous plot is used. In this type of plot a continuous surface is generated from the chromatograms given in Fig. 7.12. At the abscissa the elution volume of the SEC run (chromatograph 2) is given, whereas the ordinate gives the elution volume of the HPLC run (chromatograph 1). The peak height is assigned to a colour code, meaning that equal colours are equivalent to equal peak intensities.

The contour plot (Fig. 7.13) clearly shows three spots corresponding to the three functionality fractions, 1 being the PEG fraction, 2 being the C_{13}- and 3 being the C_{15}-alkoxy PEO fraction. It is obvious that fractions 2 and 3 have very similar molar masses, whereas the PEG is higher in molar mass.

Evaluation

The two-dimensional experiment yields separation with respect to functionality and molar mass, and the functionality type distribution and the molar mass distribution can be determined quantitatively. For calculating FTD, the relative concentration of each functionality fraction must be determined. These concen-

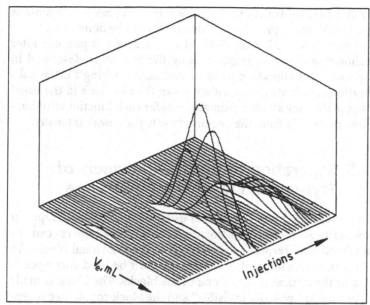

Fig. 7.12. Discontinuous plot of the two-dimensional separation of a C_{13},C_{15}-PEO sample

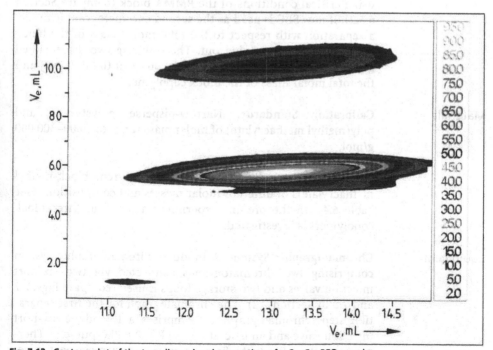

Fig. 7.13. Contour plot of the two-dimensional separation of a C_{13},C_{15}-PEO sample

trations are equivalent to the volume of each peak in the contour plot. With the appropriate software, this can be done easily.

The determination of MMD of each fraction is possible after calibrating chromatograph 2 using the procedure described in Section 7.2. Calibration must be carried out taking PEO as calibration standard. The calculation can then be done in the usual way, taking one single chromatogram for each functionality fraction, preferably from the region of the highest peak intensity.

7.5　Separation of Block Copolymers of Styrene and Methyl Methacrylate

Aim

In Section 6.5 it was shown that using liquid chromatography at the critical point of adsorption block copolymers can be analysed with respect to the length of the individual blocks. To do this, chromatographic conditions must be used corresponding to the critical point of one of the blocks. The block is made "chromatographically invisible" and the block copolymer is separated exclusively with respect to the other block.

The same principle can be used with success in two-dimensional experiments. In particular, poly(styrene-block methyl methacrylate)s can be separated with respect to the PS block using critical conditions of the PMMA block (compare Section 6.5.2). If now SEC is used as the second chromatographic mode, a separation with respect to the total molar mass of the block copolymer can be carried out. The resulting two-dimensional information then relates to the molar mass of the PS block and the total molar mass of the block copolymer.

Materials

Calibration Standards. Narrow-disperse polystyrene and poly(methyl methacrylate) of molar mass range of 1 000–300 000 g/mol.

Polymers. Anionically polymerized poly(styrene-block-methyl methacrylate)s of different molar masses and compositions (see Table 6.4). In the present experiment a blend of three block copolymers is investigated.

Equipment

Chromatographic System. A modular chromatographic system comprising two chromatographs connected via two six-port injection valves and two storage loops is used (compare Figs. 7.7 and 7.8 in Section 7.2). The chromatograph for the first separation step (chromatograph 1) comprises a Rheodyne six-port injection valve and an isocratic Type P 100 HPLC pump of Thermo Separation Products. Two electrically driven six-port injection valves (Latek Type HMV-P and ECP Type Selec-Cil) were used to connect the two chromatographs. In addition, they were connected to two storage loops of a volume of 200 µL each. The chromatograph for the second separation step (chromatograph 2) comprises a Waters model 510 pump. The operation of the coupled injection valves is directed by the software, which is used for data collection and processing. In the present case the software package "PSS-2D-GPC-Software" of Polymer Standards Service, Mainz, Germany, is used.

Columns. Chromatograph 1: LiChrospher Si-300 and Si-1000 of Merck, 10 µm average particle size and 300 and 1000 Å average pore diameter. Column size was 200x4 mm I.D.. Chromatograph 2: Four Ultrastyragel columns Linear, 1 000 and 2x10 000 Å, 10 µm average particle size and column sizes of 300x8 mm I.D.

Mobile Phase. Chromatograph 1: Methyl ethyl ketone-cyclohexane, Chromatograph 2: THF, all solvents are HPLC grade

Detectors. Evaporative light scattering detector (ELSD) model 950/40 of Zinsser Analytics in chromatograph 2.

Column Temperature. 25 °C.

Sample Concentration. 20–40 mg/mL. All samples are dissolved in the mobile phase of chromatograph 1.

Injection Volume. 50 µL.

The analysis of poly(styrene-block-methyl methacrylate)s with respect to the chain length of an individual block must be conducted at critical conditions of the complementary block (see Section 6.5). The determination of the PS blocks in the present block copolymers is carried out under chromatographic conditions, corresponding to the critical point of PMMA. The critical point of PMMA can be established on a chromatographic system comprising silica gel as the stationary phase and methyl ethyl ketone-cyclohexane as the mobile phase. As was shown in Section 6.5.2, the critical mobile phase composition is methyl ethyl ketone-cyclohexane 70:30% by volume.

Preparatory Investigations

The critical point of adsorption is very sensitive to all changes in the chromatographic system. Even using HPLC solvents from different producers may cause shifts in the critical eluent composition. Therefore, it is advisable to readjust the eluent to make sure that the system is operating at critical conditions. For the present

Separations

system a critical eluent composition of methylethylketone-cyclo-hexane 72:28% by volume was determined.

The separation of the mixture of three block copolymers (see Table 7.1), of different molar masses and composition is shown in Fig. 7.14.

This experiment is carried out at the critical point of PMMA, separating the block copolymers with respect to the PS blocks. As can be seen, separation is not very good. At an elution volume of about 3.0 mL, a peak with a shoulder at 3.5 mL is obtained, whereas at V_e=4 mL a rather symmetrical peak appears. Since separation takes place with respect to the PS blocks, it can be assumed that the first peak is due to the elution of components 1 and 2 and the second peak corresponds to component 3. In the present case, resolution of chromatograph 1 is not sufficient to separate components 1 and 2 properly.

The two-dimensional separation of the mixture is given in Fig. 7.15. Chromatograph 2 separates with respect to total molar

Table 7.1. Molar masses of the block copolymers

Sample	M_w(total) (g/mol)	M_w(PS) (g/mol)	M_w(PMMA) (g/mol)
1	182 000	93 000	89 000
2	188 000	55 000	133 000
3	20 500	9 800	10 700

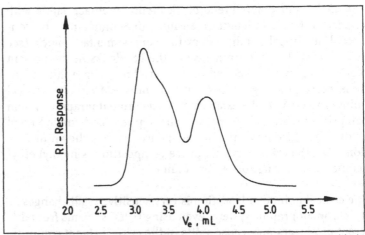

Fig. 7.14. Chromatographic separation of a mixture of three poly(styrene-block-methyl methacrylate)s in chromatograph 1, stationary phase: LiChrospher Si-300 + Si-1000, mobile phase: methyl ethyl ketone-cyclohexane 72: 28%

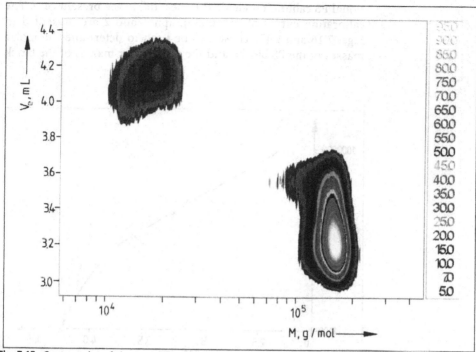

Fig. 7.15. Contour plot of the two-dimensional separation of a mixture of three poly(styrene-block-methyl methacrylate)s

mass, and therefore, the low-molar-mass component 3 appears now as a completely separated spot. Since components 1 and 2 have the same total molar mass, their separation is not improved.

In order to obtain a better separation of the three components, the efficiency of chromatograph 1 must be increased. This can be done by adding further columns of different pore sizes to this chromatographic system.

Since chromatograph 1 separates in the SEC mode with respect to the PS blocks, it can be calibrated in the conventional manner using narrow disperse PS calibration standards. For calibrating chromatograph 2 the same calibration standards are used, the calibration procedure is described in Section 7.2.

Strictly speaking, this calibration curve is not fully suitable for quantifying the block copolymers, compare discussion on SEC calibration of copolymers in Chapter 4. However, it is questionable if other calibration procedures like universal calibration or light scattering detection can be used directly in two-dimensional chromatography because of the influence of chemical composition on concentration detector response. Therefore, conven-

Evaluation

tional PS calibration yields the "best fit" in the present case. The calibration curves of chromatograph 1 and 2 are presented in Figs. 7.16 and 7.17. These can be used to determine the molar masses of the PS blocks and the total molar masses of the block

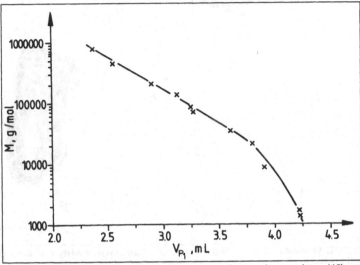

Fig. 7.16. Calibration curve of PS for chromatograph 1; stationary phase: LiChrospher Si-300 + Si-1000; mobile phase: methyl ethyl ketone-cyclohexane 72: 28% by volume

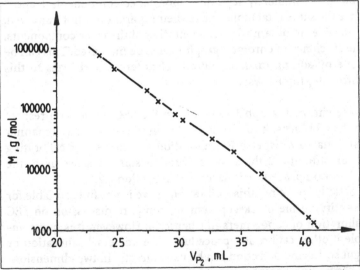

Fig. 7.17. Calibration curve of PS for chromatograph 2; stationary phase: Ultrastyragel linear + 10^3 + 2×10^4 Å; mobile phase: THF

copolymers, respectively. In the contour plot (see Fig. 7.15) the X- and Y-axes are transferred from elution volume to log M using these calibration curves.

7.6. Analysis of Styrene-Butadiene Star Polymers [12]

Unconventional polymer architectures are of increasing importance in advanced polymer systems. In addition to block and graft copolymers, star and dendrimer shaped copolymers are heterogeneous with respect to chemical composition and molar mass. Within one preparation procedure molecules with two, three or more arms can be formed, introducing a topological polydispersity. The analysis of such complex systems is complicated. SEC or adsorption chromatography alone cannot yield sufficient information on the different types of heterogeneity and, therefore, two-dimensional chromatography shall be used for the deformulation of star polymers.

Aim

Calibration Standards. Narrow-disperse polystyrene and polybutadiene

Materials

Polymers. 4-Armed star polymers based on poly(styrene-block-butadiene) were prepared by anionic polymerization according to standard procedures [14, 15] modified to give samples with well known structure and molar mass control. In a first reaction step, a poly(styrene-block-butadiene) with a reactive chain end at the butadiene was prepared. This precursor reacted with a tetrafunctional terminating agent to give a mixture of linear (of molar mass M), 2-arm (2 M), 3-arm (3 M) and 4-arm (4 M) species.

Four samples with varying butadiene content (about 20, 40, 60, 80%) were prepared in this way (see Table 7.2). A mixture of these samples is used in the present experiment. Accordingly, a complex mixture of 16 components, resulting from the combination of four

Table 7.2. Molar masses of the star copolymers

Sample	Butadiene content (wt%)	M_w (g/mol)	M_n (g/mol)
1	17	87 000	35 000
2	39	88 000	31 000
3	63	79 000	32 000
4	78	77 000	29 000

different butadiene contents and four different molar masses (M, 2 M, 3 M, 4 M) has to be separated by two-dimensional chromatography.

Equipment

Chromatographic System. A Spectra Physics 8840 pump was used for gradient formation and solvent delivery. The samples were injected by a Rheodyne 7125 manual injection valve. Transfer to the second dimension (SEC instrument) was performed by an electrically actuated Rheodyne 7010 valve equipped with a 100 µL sample loop. The injector was kept in the load position until the sample to be injected into the SEC dimension was inside the loop. Injection was software controlled by time events and contact closure. For the SEC system a Spectra Physics IsoChrom pump was used for eluent delivery. The complete system was connected to a Polymer Standards Service GPC 2000 Data Station. Data acquisition, time events and all calculations were done with 2D-CHROM hardware and software from Polymer Standards Service, Mainz, Germany.

Columns. Chromatograph 1 (HPLC): PSS non-modified silica gel, 5 µm average particle size and 60 Å average pore diameter. Column size was 300x8 mm I.D. Chromatograph 2 (SEC): Two SDV columns of PSS 1 000 and 10^5 Å, 10 µm average particle size and column sizes of 300x8 mm I.D.

Mobile Phase. Chromatograph 1: i-Octane-THF, linear gradient, 20–100% THF, Chromatograph 2: THF, all solvents are HPLC grade.

Detectors. Spectra Physics 8450 UV/vis detector and Shodex SE 61 refractometer.

Column Temperature. 35 °C.

Sample Concentration. 50–100 mg/mL.

Injection Volume. 100 µL.

Preparatory Investigations

Since the sample to be investigated is heterogeneous in different directions, the first separation step is used to separate with respect to chemical composition. The method of choice in this case is gradient HPLC [2]. The best result is obtained using bare silica gel and i-octane-THF as the mobile phase. A linear gradient starting with 20% THF in the mobile phase and going to 100% THF within the chromatographic run is used.

Initially, the 16-component star block copolymer is investigated
by SEC. As can be seen in Fig. 7.18, four peaks are obtained. They
correspond to the four molar masses of the sample consisting of
oligomers with one to four arms. The molar masses are in the
ratio M-2M-3M-4M. Despite the high resolution, the chromatogram does not give any indication of the very complex
chemical structure of the sample. Even when pure fractions with
different chemical composition are investigated, the retention
behaviour does not show significant changes as compared to the
sample mixture (see Fig. 7.18).

In each case a tetramodal molar mass distribution is visible
indicating the different topological species. The SEC separation
alone does not show any difference in chemical composition of
the samples, which vary from 20 to 80% butadiene content.

Running the sample mixture in gradient HPLC gives poorly
resolved peaks, which may suggest different composition, but
give no clear indication of different molar mass and topology
(see Fig. 7.19).

The combination of the two methods in the two-dimensional setup dramatically increases the resolution of the separation system
and gives a clear picture of the complex nature of the 16-component

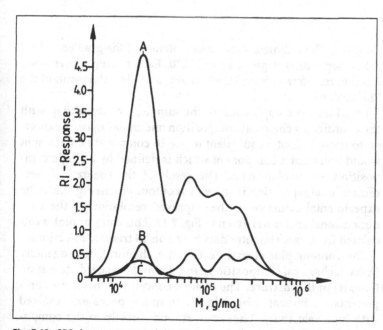

Fig. 7.18. SEC chromatograms of the 16-component sample (A) and the star poly-
mers 1 (C) and 4 (B); stationary phase: SDV PSS 1 000 and 10^5 Å; mobile phase: THF,
RI detection (Reprinted with permission from Ref. [12], Copyright 1995 American
Chemical Society)

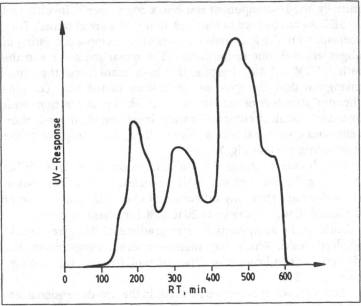

Fig. 7.19. Gradient HPLC chromatogram of the 16-component star polymer mixture; stationary phase: silica gel; mobile phase: i-octane-THF, linear, 20–100% THF (Reprinted with permission from Ref. [12], Copyright 1995 American Chemical Society)

sample. A three-dimensional representation of the gradient HPLC – SEC separation is given in Fig. 7.20. Each tracing represents a fraction transferred from HPLC to SEC and gives the result of the SEC analysis.

Based on the composition of the sample, a contour map with the coordinates chemical composition and molar mass is expected to show 16 spots, equivalent to the 16 components. Each spot would represent a component which is defined by a single composition and molar mass. The result of the theoretical two-dimensional separation is shown as contour plot in Fig. 7.21. The experimental evidence of the improved resolution in the two-dimensional analysis is given in Fig. 7.22. This contour plot is calculated from experimental data based on 28 transfer injections.

The contour plot clearly reveals the chemical heterogeneity (Y-axis, chemical composition) and the molar mass distribution (X-axis) of the mixture. The relative concentrations of the components are indicated by colours. 16 major peaks are resolved with high selectivity. These correspond directly to the components. For example, peak 1 corresponds to the component with the lowest butadiene content and the lowest molar mass (molar mass M) whereas peak 13 relates to the component with the low-

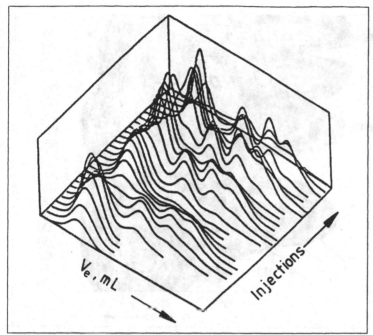

Fig. 7.20. Three-dimensional plot of the HPLC-SEC analysis of the 16-component star copolymer (Reprinted with permission from Ref. [12], Copyright 1995 American Chemical Society)

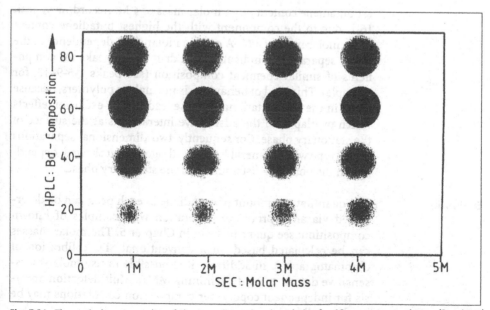

Fig. 7.21. Theoretical contour plot of the two-dimensional analysis of a 16-component mixture (Reprinted with permission from Ref. [12], Copyright 1995 American Chemical Society)

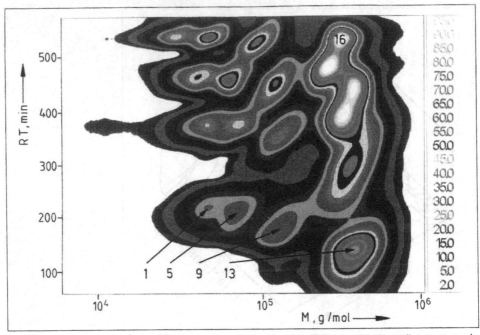

Fig. 7.22. Contour plot of the two-dimensional separation of a 16-component styrene-butadiene star copolymer (Reprinted with permission from Ref. [12], Copyright 1995 American Chemical Society)

est butadiene content but a molar mass of 4 M. Accordingly, peak 16 is due to the component with the highest butadiene content and a molar mass of 4 M. A certain molar mass dependence of the HPLC separation is indicated by a drift of the peaks for components of similar chemical composition (see peaks 1-5-9–13, for example). This kind of behaviour is normal for polymers, because pores in the HPLC stationary phase lead to size-exclusion effects which overlap with the adsorptive interactions at the surface of the stationary phase. Consequently, two-dimensional separations of this type will in general be not orthogonal but skewed, depending on the pore size distribution of the stationary phase.

Evaluation   The quantitative amount of butadiene in each peak can be determined via an appropriate calibration with samples of known composition; see quantification in Chapter 5. The molar masses can be calculated based on a conventional SEC calibration of chromatograph 2. In addition, it is possible to use molar-mass-sensitive detectors for determining MMD. Multidetection analysis for independent copolymer composition calculations may be used as well [16].

References

1. ROSENTHAL AJ, WHITE BB (1952) Ind Eng Chem 44: 2693
2. GLÖCKNER G (1991) Gradient HPLC of Copolymers and Chromatographic Cross-Fractionation. Springer, Berlin Heidelberg New York
3. BALKE ST, PATEL RD (1980) J Polym Sci B Polym Lett 18: 453
4. BALKE ST (1982) Sep Purif Methods 11: 1
5. GLÖCKNER G, VAN DEN BERG JHM, MEIJERINK NL, SCHOLTE TG (1986) In: Kleintjens I, Lemstra P (eds) Integration of Fundamental Polymer Science and Technology. Elsevier Applied Science, Barking
6. GLÖCKNER G, STICKLER M, WUNDERLICH W (1989) J Appl Polym Sci 37: 3147
7. AUGENSTEIN M, STICKLER M (1990) Makromol Chem 191: 415
8. MORI S (1990) J Chromatogr 503: 411
9. MORI S (1988) Anal Chem 60: 1125
10. MORI S (1981) Anal Chem 53: 1813
11. KILZ P (1993) Labor Praxis 6: 64
12. KILZ P, KRÜGER R-P, MUCH H, SCHULZ G (1995) ACS Adv Chem 247: 223
13. KILZ P, KRÜGER R-P, MUCH H, SCHULZ G (1993) PMSE Preprints 69: 114
14. SCWARC M, LEVY M, MILKOVICH R (1956) J Am Chem Soc 78: 2656
15. CORBIN N, PRUDHOMME J (1976) J Polym Sci Polym Chem Ed 14: 1645
16. KILZ P, GORES F (1993) In: Provder T(ed) Chromatography of Polymers. ACS Symp Ser 521. ACS, Washington, Chapter 10

References

Subject Index

Springer
and the
environment

At Springer we firmly believe that an
international science publisher has a
special obligation to the environment,
and our corporate policies consistently
reflect this conviction.

We also expect our business partners –
paper mills, printers, packaging
manufacturers, etc. – to commit
themselves to using materials and
production processes that do not harm
the environment. The paper in this
book is made from low- or no-chlorine
pulp and is acid free, in conformance
with international standards for paper
permanency.

 Springer